THE EFFORTLESS SOUS VIDE COOKBOOK

the *the* effortless sous vide COOKBOOK

140 RECIPES FOR CRAFTING RESTAURANT-QUALITY MEALS EVERY DAY

BY CAREY COPELING

Photographs by Marija Vidal

ROCKRIDGE
PRESS

For general information on our other products and services or to obtain technical support, please contact our Customer Care Department within the United States at (866) 744-2665, or outside the United States at (510) 253-0500.

Rockridge Press publishes its books in a variety of electronic and print formats. Some content that appears in print may not be available in electronic books, and vice versa.

TRADEMARKS: Rockridge Press and the Rockridge Press logo are trademarks or registered trademarks of Callisto Media Inc. and/or its affiliates, in the United States and other countries, and may not be used without written permission. All other trademarks are the property of their respective owners. Rockridge Press is not associated with any product or vendor mentioned in this book.

Photography © Marija Vidal

ISBN: Print 978-1-62315-981-8 | eBook 978-1-62315-982-5

contents

introduction
Sous Vide Time

Toast and takeout: That was the extent of my ability to provide basic sustenance for myself. In my youth, I was fortunate to have been fed extremely well by my highly talented mother, but after my first year in college, when the student residence meal plan ended, I was on my own. I ate poorly, without concern for nutrition, and I definitely was not a good cook.

Fast-forward 16 years, to when Chris, a friend and client of mine, posted on Facebook, "If you like meat and fish, get an Anova sous vide—it will change your life." I was intrigued and within a few clicks had added it to my cart and clicked "checkout." Initially, the sous vide was a gift for my then girlfriend, now wife, Mel, who was visiting her family in Sydney, Australia. When it arrived, my intrigue grew into childish curiosity and then became full-blown impatience. I couldn't wait for Mel to get back from her trip. I popped open the box, quickly justifying it by thinking that if I figured out how it worked, it would save some initial frustration that often comes with learning how to use a new kitchen gadget.

I always did enjoy grilling—especially steak—and still do. After some quick research on the basics of sous vide cooking, I purchased a butcher-cut New York strip loin steak. Following a basic recipe, I liberally salted and peppered the beef, slipped it into a large resealable bag, and after a few attempts managed to get the air out of the bag using the water displacement method (don't worry, we'll come back to that later). I prefer medium rare, so I set the temperature at 129°F (53.8°C) and one hour and thirty minutes later, out came the steak. But this wasn't just any steak; it was *the steak*, the best steak I'd ever made. I was hooked, and it was easy.

By the time Mel returned from Australia, I had cooked a dozen different meals—chicken breast, asparagus, self-buttering corn on the cob, octopus, a rack of lamb, just to name a few—in a week. I'm not sure if Mel knew it yet, but she came home to a changed man. In a few short weeks, I was cranking out really, really good meals. I was hosting dinner parties—yes, me, Mr. Toast and Takeout—and wowing my friends and family. One of my favorite things about cooking sous vide is that after you've prepared the dish and popped it in the water bath, you're able to join your guests and enjoy their company, confident that your creation is on track, cooking perfectly, and will turn out just as you expect. Yes, some foods take longer to cook sous vide, but what you get back is time with your family, your kids, to listen to music, read a book, enjoy a glass of 2007 Rioja, or whatever you want to do with your sous vide time while your meals cook to perfection.

Many of you are familiar with sous vide cooking, so you likely know that chefs all around the world have been using this cooking method for decades because it helps them manage large volumes of main or side dishes and variations of doneness. It also involves a less hands-on/eyes-on approach, which allows them to focus on some of the finer details that make up your meal. Until a few years ago, in-home sous vide appliances were cost-prohibitive, required large amounts of surface or counter space, and didn't come with any instructions on how to use them. Today, there are dozens of sous vide machines specifically designed for the home chef.

In this book, we are going to cover some adventurous recipes, but what really inspired me the most in the kitchen was taking the basics of a dish and making it my own. I love to travel, so I'm exposed to cultural cuisines that, both literally and figuratively, get my creative juices flowing. I try to incorporate that into my sous vide recipes. I encourage you to experiment with your own recipes, mix in some cultural nuances, perhaps even tweak Grandma's secret recipes, and turn basic dishes into something incredibly delicious.

chapter one

Fine Dining Every Day

If you're reading this book, you've likely purchased a sous vide circulator and you're wondering, "How do I get the most out of this thing?" We are going to cover some basics—equipment, methods, and safety measures, to name a few—but this book is really focused on pushing your sous vide skills further. You'll try some unique foods, some longer and more advanced recipes, and some new techniques to help you discover just how amazing sous vide cooking can be. The sous vide community grows larger every day, and with it will come new products, new recipes, and new applications for sous vide cooking. I've learned a lot in the few years since *the steak* and encourage you to participate and collaborate with others to share and learn from each other's experiences, both successes and failures. I hope this book inspires you to be creative and innovative when it comes to cooking that perfect meal. Bon appétit!

Time Well Spent

Even after cooking hundreds of meals using the sous vide method, there are things I'm still learning today. The range of foods you can cook is endless, but I did come up with a few other reasons why sous vide cooking is more rewarding than some of the more traditional methods of cooking. Here are just a few.

ACCURACY   Precision cooking is exactly that: cooking a meal or component to an exact temperature. Looking at it another way, it removes the guesswork associated with cooking, especially when you are trying something for the first time. My best example is from my own experience of cooking octopus, one of the most difficult foods to cook correctly. If you cook octopus 5 minutes too long, or 5 degrees too hot, you'll miss the texture you were aiming for. I didn't. I nailed it on the first try.

PREDICTABILITY   Despite variations of a particular type of meat or veggie (think fat percentage, or length/girth), you can consistently deliver the same outcome over and over again. Conversely, if cooking via a traditional method—oven or grill—you can lose heat by opening the lid or door, which can result in an increase in cook time and moving the goalposts of a perfectly cooked meal.

RANGE   Every day, it seems, the sous vide community comes up with some-thing new that can be cooked using sous vide methods. The variety of foods and drinks that can be transformed appears to be endless. I've been able to cook several entire meals—every dish—using my sous vide tools, including a cocktail, appetizer, main course (with veggies or potatoes), and a perfect dessert. That's hard to beat in terms of the spectrum of things you can do.

TIME   While cooking sous vide does typically take longer than other tradi-tional methods, it allows you to better use your time. The preparation work for sous vide versus traditional cooking methods is roughly the same; there aren't any special steps you need to do to get your meal ready to take a swim in the water bath apart from the vacuum seal, or equivalent, right before you cook. It's the worry-free aspect of the sous vide time that is the real advantage. You know exactly when the food is ready, and if you aren't ready, almost all recipes will be fine if you let them sit for another 10, 30, or sometimes 60 minutes or more. Keep your cocktail hour going if the conversation is good—the food won't overcook or spoil.

ADVENTURE   After a few basic dishes, you're going to want to push your limits. I mentioned octopus earlier, a notoriously difficult dish to master. It was one of the first ten things I ever cooked sous vide. Before then, I'm not sure I could recall seeing an octopus outside of a Mediterranean restaurant in my life, but afterward, I could have octopus any time I wanted. And while not every dish has to be something exotic, sous vide can awaken your culinary adventurous side because dishes that were previously outside the realm of your possibilities are suddenly within reach.

PORTIONING   You can cook for one or cook for one dozen and be confident that the whole group will have the same experience. Having said that, there are limits to how many steaks, for example, you'll be able to manage with a single home-based sous vide machine, as each circulator can only maintain precise temperatures for a certain volume of water. While you wouldn't be able to keep 100 pieces of beef at 129°F in your bathtub, you could cook more—and more consistently—than you would have been able to without your sous vide machine. I once cooked a six-pound beef tenderloin for a group of 14 people using two Anova sous vide machines and a Rubbermaid® bin, so you can expand your capabilities if you have the means to do so (i.e., more than one machine). Anyone who has tried to cook a large meal for a large group knows that it can be daunting. It typically involves you spending the entire evening in the kitchen trying to coordinate all the complexities and components, but with sous vide and some planning, you'll be able to manage your dinner parties stress-free.

WOW FACTOR   The flavors, the science, the tech, the edge-to-edge medium rare—they all contribute to a very impressive meal. If your conversation before dinner wasn't about sous vide cooking, it will be after your guests take their first few bites. Chicken breast cooked entirely on the grill versus those cooked sous vide and finished on the grill might *look* similar, but once you cut into it, the difference is immediately apparent. The texture, the flavor, the moistness, the "you took something that is dry and made it melt in my mouth" all combine to blow the collective minds of the people around your table. When was the last time you were able to do that?

sous vide then and now: a little look back

- In 1799, Benjamin Thompson, Count Rumford, was attempting to cook mutton by precisely maintaining a specific temperature of 212°F (he was using air, not water, and invented the convection oven). After three hours, he checked his meat and was disappointed to find it wasn't nearly cooked. Frustrated, he stormed out of the kitchen and left the mutton to be cleaned up by the kitchen staff the next morning. To the staff's surprise, not only was the mutton cooked when they checked in the morning, it was delicious.
- The idea of low-temperature cooking lay dormant until 1960, when French and American engineers began manufacturing food-safe plastic bags and films for Cryovac to seal and pasteurize industrial trial foods for longer shelf life.
- Frenchman Pierre Troisgros discovered in 1974 that cooking foie gras sous vide resulted in very little fat and weight loss, maintaining its original shape and texture.
- In the early 1990s, sous vide cooking in France was considered only fit for industrial food, and a television report revealing that Chez Margot was cooking using sous vide methods caused the restaurant to go bankrupt soon after.
- In the mid-2000s, sous vide cooking became widely known and appreciated for its consistency and ability to maintain fat and moisture lost through traditional cooking methods.
- Nowadays, there are over 100 different in-home sous vide circulators available, with more being kick-started every month. The lowered costs and adaptation of mobile phone technology has made this historic cooking method accessible and approachable for all levels of cooks

Getting the Most Out of This Book

I've broken out the recipes by food type (poultry, pork, vegetables, etc.) versus courses (appetizers, mains, etc.) because the components related to the food type can be used across an entire four- or five-course meal. For example, chicken breasts can be used in salads, in soups, as a main course, and even as a snack.

When trying to determine timing, temperatures, and weights, it can be challenging to understand whether or not something has been cooked "enough" to be safe, or long enough to break down connective tissues, or the impact thickness can have on timing. I provide the weights, temperatures, and times based on my tested recipes, although you can make adjustments based on your tastes. I've also included what is considered a best practice guide for cooking times and temperatures for the most commonly used foods (page 199).

You'll see that some of the first recipes in this cookbook aren't actually sous vide recipes. They are a combination of rubs, marinades, and sauces that I use as flavor profiles throughout the various chapters. Understanding how different flavors impact different ingredients will help you appreciate more subtle flavors versus the sometimes-overpowering salt or pepper flavors typically associated with more basic recipes. The complexity of the flavors is what makes an ordinary dish extraordinary. With sous vide cooking, those flavors are supercharged when locked and sealed in with your foods.

Essential Kitchen Equipment

Apart from the obvious (hint: it's a sous vide machine), there are several additional pieces of equipment that will help you get the most out of your recipes. Some are optional, while some are essential and will have an impact on the quality, safety, and overall success of your meals.

FOOD-SAFE SOUS VIDE BAGS

First and most important are your cooking bags. The bag is the only layer that comes into direct contact with the food, so you need to make sure they are food-safe. You can determine this from the material used in their manufacture.

The materials that are considered food-safe, and the best to use safely with sous vide cooking are made from:

- Food-safe high-density polyethylene

- Food-safe low-density polyethylene

- Polypropylene

    In terms of where and how to get them, I recommend the following options:

- Glad Ziploc® brand bags are considered food-safe and perform well if you don't have a vacuum or chamber sealer (see page 10 for the differences).

- FoodSaver® brand bags are also considered food-safe and can be purchased with a vacuum sealer or as refills once the initial allotment that comes with the machine runs out.

- Amazon sells off-brand vacuum-seal bags at a fraction of the cost of the FoodSaver® bags that work with most vacuum sealers. You'll need to read the details and reviews to determine their manufacturing material and safety for use with food.

- A few products offer reusable bags and use a hand-pump to remove the air from the bag. They have a press-to-seal top, similar to Ziploc®, and a small flap in one corner where the hand-pump is used to suck out the air. I have used a kit made by For Good, and the bags can be reused three or four times before they start to lose their seal and shape. These can be handy at times when you don't have a vacuum or chamber sealer.

    An alternative to cooking bags is a glass mason, canning, or preserves jar. Apart from the obvious reduction in environmental impact, glass jars are useful for specific recipes such as infusions, desserts, and egg bites, to name a few. They can be washed and reused, which is economically effective. You can't use glass jars for most recipes, but it's good to have some available to be used when possible.

WATER CONTAINERS

Sous vide cooking temperatures can range from 100°F to 200°F; rarely do recipes call for setting the water temperature over 200°F. Sous vide recipes also

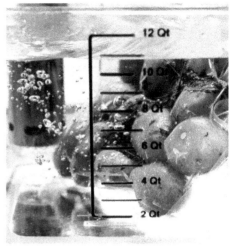

L: PROPER PLACEMENT OF FOOD IN VACUUM SEALER-STYLE BAG. R: VACUUM-SEALED POTATOES UNDERWATER.

never call for boiling water, but recipes using higher temperatures could result in warping, cracking, or potentially melting poorly or improperly made containers. Here are some proven options for containers.

- My personal favorite is the square, 12-quart/liter clear Rubbermaid® food storage container. It is food-safe, used in restaurant kitchens for safe food storage, and has the depth required for minimum water levels across all sous vide machines. Its snap-tight lid—I've cut out one corner of mine to minimize evaporation within the container—is sold separately. The container is light, easy to carry, and, because it is clear, provides a window to the food while it cooks.

- If you don't wish to buy a new container, you might have a saucepan deep enough to meet the minimum water level requirements. A bonus to using a saucepan is that you can add heat from your stovetop burner to heat the water much faster than just using the sous vide machine's heating abilities. Double bonus: If you have a single sous vide machine and you want to cook veggies after cooking a protein low and slow, you can use the elemental heat to quickly go from, say 140°F water for chicken to 185°F water to quickly cook asparagus. This trick can save some crucial minutes if you're attempting the two-recipe-one-machine combinations.

○ Some home chefs convert a medium-size cooler into an insulated sous vide container to use for recipes with longer cook times. While requiring less energy to keep the water warm, some of that efficiency is lost if you have to remove the lid to insert the circulator. There are many guides online on how to bore out the perfect-size hole in the lid for various sous vide machines. Once you've done it, however, you can only use that cooler for sous vide and not much else.

CAST IRON PANS, GRILLS, AND TORCHES

Proteins, particularly chicken or beef, don't look their most appetizing when removed from the water bath. Many recipes call for a good hot sear to add a crust, crunch, or smoky flavor. There are a few ways to achieve this. The most important thing is to expose the food to a very hot source so it sears fast without overcooking. How hot? My rule of thumb is a minimum of 500°F. I've had success with the following methods.

Invest in a good **cast iron pan.** You'll get more use out of it than just finishing or searing sous vide dishes, and it can withstand more heat than a traditional sauté pan. In fact, Teflon®-coated pans are a no-go for searing sous vide dishes; the nonstick coating can burn off and enter the food. Simple rule: If you're going to use a pan to sear, make sure it's cast iron, no exceptions. I chose the word "invest" because a properly seasoned cast iron pan can last you a lifetime. Check out the Resources (page 202) for a link on how to properly season your pan for first use and subsequent maintenance. See the section on oil and smoking points; the one downside to cast iron pans is the smoke and spit, which can quickly set off your smoke detector.

If weather permits, you can use your outdoor **grill** (unless you have an indoor grill that meets the minimum 500°F temperature requirement) to finish your food. Some new grills, such as a few Weber models, have a dedicated searing function, or a searing station. These elevated burners help you reach a much higher temperature in a specific area of your grill and are ideal for finishing red meats and poultry. With temperatures reaching between 500°F and 800°F, you won't need to sear for very long, and remember, you are not finishing the cooking, you're searing. Once you have some good crust or grill marks, remove the meat immediately, otherwise you'll end up ruining the low and slow cooking you just patiently waited for.

FINISHING YOUR COOKS: PAN SEARING (L), WITH A TORCH (R)

If you don't have a grill (hello, apartment dwellers) and you suffer from poor kitchen ventilation, you still have an option: the **torch.** Using a strong propane torch allows you to sear most dishes; it just takes time. Not all torches are the same, however. Smaller, less expensive torches that are typically used for desserts (think crème brûlée) won't do you much good as they don't get hot enough to generate a meaningful sear. The most effective torches are ones that attach directly to a small propane tank. Two brands I've used are Bernzomatic® and Mag-Torch. Each offers a push-button starter and generates plenty of heat (and very little smoke) to finish the protein with a good sear. There are additional attachments that can be used with propane torches to help displace the flame over a wider area. I used one such attachment, Searzall, but found that it worked better on some dishes (fish) but not as well on others (beef). Don't be fooled into thinking you must have a displacement attachment; a good solid torch will do the trick.

WATER DISPLACEMENT METHOD VS. VACUUM SEALER VS. CHAMBER SEALER

While this section is focused on essential kitchen equipment, I just want you to know that you don't have to use a vacuum sealer or a chamber sealer to be successful with sous vide cooking.

The **water displacement method** is an excellent alternative to the expensive bags, potential increased waste of vacuum/chamber cooking bags, and investment required for a vacuum or chamber sealer. Given that you are working with a vessel full of water, the water displacement method naturally fits with sous vide cooking. Place the food inside a resealable cooking bag and seal from one side to the other, leaving the last ½ inch of the bag open. Using a spatula (especially if the water is hot), start pushing the food and the bag down under the water. The pressure of the water on the bag will expel the air, creating a tighter fit around the food. Continue to lower the food into the water until the water level approaches the zipper then close the last ½ inch of the bag and remove the bag from the water. You should have a seal that is very close to airtight. For most dishes, this method will allow the food to sink and remain below the surface of the water. The one exception I've found is with vegetables—they almost always float. Even with a vacuum-sealed bag, keeping vegetables beneath the water surface is a challenge. Lay a dinner plate on top of the bag to provide the extra weight needed to keep the bag underwater for the duration of the cook.

The next step up in terms of getting a nice tight vacuum around the food is to use a **vacuum sealer.** Most vacuum sealers are clamshell sealers, meaning you place the food-safe vacuum bag under the lid, close, and seal. These sealers remove most of the air from the bag and then use heat to seal the bag. They are more effective in terms of removing air although they can be somewhat violent as the food is squeezed to remove the air. Delicate foods like fish can lose their shape or show pressure marks where the bags were pulled tight. Another challenge with vacuum sealers is that they don't do very well at sealing foods in liquid. The sealer will often suck up the liquid while vacuuming out the air, which can cause the seal to fail. If you drape the bag over the edge of the countertop while sealing, you can reduce the risk of that happening. I've almost always used a vacuum sealer to prepare my food for a sous vide water bath, and I invested in a sealer that has a manual stop-and-seal function so I can

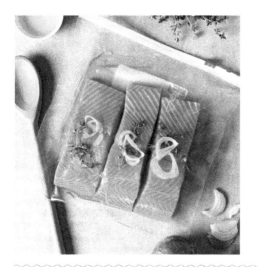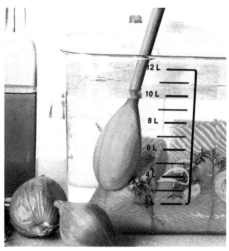

RESEALABLE BAGS: PREP AND SUBMERGING INTO A BATH.

control the amount of vacuum on a dish-by-dish basis. When you are submerging food underwater, you want to make sure the seal holds. Some manufacturers are better than others, so do your research and read customer reviews; they will help you understand which sealers work best. Prices for a decent sealer start at about $99.

The **chamber sealer** is what I call the professional option. It's definitely on my wish list, but it is very expensive compared to vacuum sealers, and even more so compared to the water displacement method. With a chamber sealer, the entire bag is placed inside the sealer then the lid is brought down and closed tight. All of the air is removed in the chamber before a heated sealing bar is brought down over the open edge of the bag. The biggest benefit of a chamber sealer is that it can remove the air and seal just about anything, including liquids. As I mentioned, chamber sealers are professional grade and often used in butchers, grocery stores, and other food suppliers. They start at about $750 and can go for more than $3,000.

There are a few areas where we can start to "score" the machines to find the one that best suits you, and I'll leverage my marketing analytics experience to help qualify and quantify based on those parameters.

Sous Vide Circulators: Compare and Contrast

What was once a barren sous vide machine market has now exploded. Everyone likes choice, but with these devices, they all claim to do the same thing, they all have roughly the same shape, and they all cost roughly the same amount. How do you decide which one is best for you? The following table covers what the main machines on the market have to offer.

WI-FI: IS IT WORTH IT?

One of the biggest differentiators and cost considerations is whether to splurge on a circulator that includes a Wi-Fi option. Almost all new sous vide machines offer a Wi-Fi option, which allows you to remotely control and monitor your sous vide machine from outside the home. You can turn it on, change temperature, and detect any issues with water levels or power outages, for example, and this functionality is designed to help you have your food ready, or close to, by mealtime. I've owned a few Wi-Fi–enabled sous vide circulators, and I've only used the function a handful of times. One challenge is that you have to plan your meal before you leave, calculate the time it takes to cook, and work backward from that to determine when to turn it on. Probably the biggest challenge is keeping the food cold and fresh until you turn on the machine. You can't leave meat, for example, sitting in a warm water bath all day: That's dangerous. So you need to keep the water cold until you're ready to start the cook. It is possible to do, and my sister, Nicki, does it often. You have to fill the container with a mixture of water and ice to maintain a safe temperature. The ice will melt slowly throughout the day (you'll need to play around with the melting time before attempting a recipe) until you start your machine, at which time the ice will melt very quickly and your food will begin to cook. You'd need to factor in additional time to get to the correct cooking temperature. The cook time starts once the water reaches the appropriate cooking temperature. If you have a very active and busy life, spending the extra $50 to $75 for a Wi-Fi version might be worth it.

With Anova, you can locate a recipe, pair your device with your smartphone, and begin your cook. It's really that easy. With the ChefSteps Joule, you can do the same thing—in fact, with the Joule, you can only control your device via the app as there is no way to manually change the temperature of the device without using your phone. Sansaire™'s device, on the other hand, has no connectivity to any app or device, so you must manually control time and temperature on the device itself. It really comes down to personal preference, but I like the devices that can connect to my phone, have a good catalog of recipes, but also allow for manual controls when I want to go "off book."

PUMPS AND DEPTHS

I know many beginners who have run into this problem: They have just purchased their first sous vide. They are ready to cook, but their pot isn't deep enough to meet the minimum immersion requirements of their new device. So frustrating! Now they have to wait a few more days to get a container delivered before they can enjoy great sous vide meals. Anova is the biggest culprit since it requires quite a lot of water height to run safely (7.25 inches, to be exact) whereas Joule, because it pulls from the bottom of the circulator, can run in 1.5 inches of water. In fact, most of the new sous vide machines can run in less than 4 inches of water, which means you can likely use a pot you already own. Be conscientious of your machine's requirements and make sure you have an appropriate container.

Lastly, when determining how many servings (bags) you can circulate at once, you need to consider the volume of water that can be circulated by the pump. This is measured in terms of how much water your sous vide can circulate per minute. Combining this with the heating power of the machine will help you calculate how quickly you can get your water bath up to temperature. As an example, the Sansaire™ circulator has 1100 watts of heating power and can circulate 11 liters (3 gallons) per minute. Compare that to Anova's circulator, which has 800 watts of power and can circulate 8 to 9 liters (2.1 to 2.4 gallons) per minute. This means that if you need to heat water quickly, the Sansaire™ will get to your desired temperature faster.

MODEL	MAXIMUM TEMPERATURE	HEATING	PUMP	POWER AND SAFETY
Anova Precision Cooker	77°F (25°C) to 210°F (99°C)	800W	2.1 to 2.4 gallons per minute 8 to 9 liters per minute	Power input: 110–120VAC (220–240V EU) Bi-metal fuse
ChefSteps Joule	68°F (20°C) to 208°F (98°C)	1100W	1.6 gallons per minute 7 liters per minute	Power input: 110 to 120VAC Frequency: 50 to 60Hz, single phase Secure data transfer over Wi-Fi and Bluetooth Smart Thermal cutoff protection High current protection
Sansaire™	32°F (0°C) to 212°F (100°C)	1100W	3 gallons per minute 11 liters per minute	Power input: 110V (220 to 240V EU) Frequency: 60Hz Overheat / low water level protection
Nomiku	32°F (0°C) to 203°F (95°C)	1150W	1.6 gallons per minute 7 liters per minute	Power input: 110V (220 to 240V EU) Frequency: 60Hz Conforms to UL STD 1026 safety standards Power Loss Warning Bacteria Growth Warning
Gourmia GSV150	32°F (0°C) to 203°F (95°C)	1200W	1.5 gallons per minute 5.7 liters per minute	Power input: 120V Frequency: 60Hz ETL marked of safety standards Auto shutoff

DISPLAY AND CONTROLS	CONNECTIVITY	WEIGHT AND DIMENSIONS	COST
LED Display Set timer for up to 99 hours LED backlit scroll wheel	Bluetooth: BLE 4.0 Wi-Fi: BLE 4.0 and 802.11 Wi-Fi iOS app Android app	Length: 2.75 inches / 6.98 cm Width: 2.75 inches / 6.98 cm Height: 14.75 inches / 36.8 cm Immersion Depth: 7.25 inches / 18.415 cm Max clamp opening: 1.25 inches / 3.175 cm Minimum water depth: 2.5 inches / 6.35 cm Weight: 2.5 pounds / 1.1 kg	$149 (Bluetooth) $199 (Wi-Fi + Bluetooth)
App display only	Bluetooth Smart 4.0 wireless technology IEEE 802.11 b/g/n Wi-Fi compatible	Length: 1.85 inches / 4.7 cm Width: 1.85 inches / 4.7 cm Height: 11 inches / 28 cm Max clamp opening: 0.75 inches / 1.9 cm Minimum water depth: 1.5 inches / 3.8 cm Maximum water depth: 8 inches (20.3cm) Weight 1.28 pounds (580 g)	$199
LED Display Twist to change temperature	No wireless connections or apps	Length: 4 inches / 10 cm Width: 4 inches / 10 cm Height: 15 inches / 38 cm Minimum water depth: 2.75 inches / 7 cm Maximum water depth: 6.5 inches / 16.5 cm Weight: 4 pounds / 1.8 kg	$199
OLED Display Twist to change temperature	Eat Tender App (Wi-Fi Only) IEEE 802.11 b/g/n Wi-Fi compatible	Length: 6.5 inches / 16.5 cm Width: 6.5 inches / 16.5 cm Height: 13.5 inches / 34.3 cm Minimum water depth: 3.5 inches / 8.9 cm Maximum water depth: 6 inches / 15.24 cm Weight: 2.75 pounds / 1.2 kg	$199
LED Display LED backlit scroll wheel	Wi-Fi: and 802.11 Wi-Fi iOS app Android app	Length: 3.5 inches / 8.9 cm Width: 3.5 inches / 8.9 cm Height: 14.5 inches / 36.8 cm Minimum water depth: 2.75 inches / 7 cm Maximum water depth: 4.8 inches / 12.4 cm Weight: 2.3 pounds / 1.0 kg	$249

don't sweat the technique

The first time you cook sous vide, you'll likely have more questions than answers, but don't worry, here are a few tips to help you get the most out of your cooks.

BEFORE YOU BAG

Seasoning: It is especially important to season protein *before* putting it in the bag. If you sprinkle the seasonings into the bag, you will end up with uneven seasoning and perhaps too much or not enough in parts of the dish.

Hot Sauces and Marinades: Let cooked sauces and marinades cool for 5 to 10 minutes before adding to the bag, to avoid pouring near-boiling liquid onto uncooked protein.

DON'T SQUEEZE TOO HARD!

One of the biggest benefits of the water displacement method versus vacuum or chamber sealers is its gentler handling of delicate foods such as fish. Many lower-end vacuum sealers have an all-or-nothing approach to the way they seal, with a handful of higher-end sealers providing a manual option. If your vacuum sealer doesn't have a manual seal feature, consider using the water displacement method to ensure you don't crush delicate foods. Foods like fish or foie gras can lose their shape very easily under the pressure of a vacuum sealer, so don't be afraid to switch between vacuum and water displacement methods as needed. Try to ensure that you don't crush, break, or squeeze the food so it retains its shape.

DON'T OVER SEAR

You just allowed extra time for your dish to cook low and slow. The last thing you want to do is leave a perfectly medium-rare steak in a hot pan, or on a hot grill, for too long and discover you've cooked it all the way through. Because the food is warm (and already cooked—remember that) heat will transfer much more quickly to the food than if it was just taken from the refrigerator. Your best bet is to get the grill or pan as hot as you can get it so the food spends less time searing. See Cast Iron Pans, Grills, and Torches (page 8) for more details on searing.

YOU'VE SPRUNG A LEAK

If water leaks into the bag while you're cooking, more often than not the food will be spoiled. This most often occurs during longer cooks when the bags are exposed to heat for a longer period of time. Don't try to salvage a waterlogged steak; learn from your mistake and double-bag and double-seal your foods if you're worried. There is no harm in double-wrapping; the heat will pass straight through and you'll be covered in case one of the bags springs a leak.

units of measurement

Depending on where you live in the world, you'll use Fahrenheit or Celsius—there is no right or wrong choice. You might be able to convert some temperatures in your head, but for the most part you probably have to look up any conversion from °F to °C. We'll do you the favor of providing the conversions for the recipes in this book. Most sous vide circulators support setting temperatures to 0.1 of a degree (F or C) so you can be very precise. If for some reason you are only able to set whole-digit temperatures, set your machine to Fahrenheit as the differences in a single degree of temperature is less than with Celsius.

In the same way, you're bound to find that different recipes use different units of measurement for ingredients. Some recipes will use measurements of weight (grams, ounces), others by dry volume (tablespoons, cups), and still others in more obscure units (sprigs, thumbs). You can't be expected to know them all! My solution is to use an app on my phone—it converts temperature, measurements, units, etc. really quickly, and most apps are free. I use the All-In-One Calculator (I'm an Android user), which includes all cooking-related calculations.

Time to Get Cooking

Ready to get started? Here are a few things to keep an eye on to make sure you have a successful cook.

NO SPLASHING!

While you never "boil food in a bag" (a surprisingly common misconception about sous vide cooking) you are working with hot water. Be careful when lowering bagged food into the water; don't just drop it into the middle of the water. Your best bet to avoid any backsplash is to slide the food down the side.

This is the perfect time to fasten the bag to the container's side with a binder clip, so the bag doesn't circle around the machine and impact the water flow. Pinning the food to the container side also makes retrieval easy and safe. Silicon or rubber oven mitts can be used to retrieve bags that slip to the bottom of the water bath.

THREE IS A CROWD

When you are cooking multiple large proteins at the same time, don't add too many pieces to a single bag. My basic rule of thumb is no more than two protein pieces per bag. Any more than that and the pieces will always be touching each other, which means there will be protein surfaces not enveloped by the water, which means there's a risk of uneven cooking. It's better to use two or more bags, as needed. A definite no-no is stacking proteins. Meat or fish will definitely be undercooked. Always cook sous vide with the food in a single layer.

FIND SOUS VIDE FRIENDS

Of course, you'll have friends that also sous vide, but that's not what I mean. As you move through these chapters, you'll note that there are certain foods that cook at the same, or very close, temperatures. This is a great way to double up on the opportunity to sous vide complementary dishes. While the cook times might be different, you can use the same bath for two dish components. Calculate the differences in cook time and start with the longer cook first, adding the shorter cook food later so they both finish at or around the same time. An easy example to try: Almost all the fresh vegetable recipes in this cookbook cook at 185°F (85°C) so you can cook a variety at the same time (in separate bags) and have mixed vegetables all ready together!

chapter two

Basics, Sauces, Spices, and Marinades

The foundation of any excellent sous vide dish is the flavors included with the fish, veggies, or meats. The most basic recipes often call for just salt and pepper, but once you've mastered that, you're going to want to experiment with more complexity to unlock restaurant-quality dishes. I've compiled some my favorites in the form of rubs, sauces, spices, and marinades to help get you started. I use and reference these throughout the book, which will give you an idea of how you can take traditional cooking sauces and rubs and incorporate them into your sous vide recipes.

POACHED EGGS TWO WAYS (page 39), HARD-BOILED EGGS (page 40), SOFT-BOILED EGGS (page 41)

DRIED GREEN CHIMICHURRI RUB *and* MARINADE

Prep time: 10 MINUTES

This versatile rub can be used on everything from beef and lamb to fish and poultry. It can also be turned into a marinade by adding oil and vinegar. Chimichurri is typically bright green and has South American roots. Loosely translated, chimichurri means "a mixture of several things in no particular order," making it a fun way to spice up your dishes. **¼ CUP**

FOR THE RUB

3 tablespoons dried basil leaves
3 tablespoons dried oregano leaves
2 tablespoons dried parsley flakes
2 tablespoons kosher salt
2 tablespoons dried thyme leaves
1 tablespoon freshly ground
  black pepper
1 tablespoon paprika
2 teaspoons garlic powder
1 teaspoon red pepper flakes

FOR THE MARINADE (OPTIONAL)

½ cup olive oil
⅓ cup red wine vinegar

TO MAKE THE RUB

Mix the basil, oregano, parsley, salt, thyme, pepper, paprika, garlic powder, and red pepper flakes in a medium bowl and rub liberally on the protein (steaks, lamb, whitefish, or chicken) before sealing in a bag. Store any leftover rub in an airtight container for 2 to 3 weeks.

TO MAKE THE MARINADE

Add the oil and vinegar to the rub in a medium bowl, whisk together, and serve immediately. You can also store the marinade in the refrigerator, and use within 2 to 3 days.

FAT PIG (*FETTE SAU*) DRY RUB

Prep time: 10 MINUTES

This is my favorite pork rub, and I encourage you to make any adjustments you need to get the flavor and texture that will make your taste buds sing. When preparing larger portions of pork—think pulled pork or ribs—you'll need a large amount of rub to cover the front, back, and sides. This recipe should make enough rub for most dishes, and you can always store the rest in an airtight container and use within 2 to 3 weeks. **3½ CUPS**

1¼ cups packed dark brown sugar
1 cup kosher salt
¾ cup finely ground espresso beans
⅓ cup garlic powder

¼ cup freshly ground black pepper
2 tablespoons ground cinnamon
1 tablespoon cayenne pepper
1 tablespoon ground cumin

Thoroughly mix all the ingredients in a bowl and rub liberally on the protein (steaks, ribs, pork shoulder/butt, chicken legs) and seal in a bag.

Tip: If you are looking for a coffee substitution, you can use finely ground cocoa beans or cocoa powder.

CHIPOTLE CHICKEN RUB

Prep time: 10 MINUTES

This super simple chipotle rub is perfect for adding a fiery touch to your chicken dishes. The recipe has only four ingredients, all easily found at your grocery store. Chipotle chiles are smoke-dried jalapeños but are not overly hot, so guests with milder palates will get the spice without too much burn. **3½ CUPS**

¾ cup chili powder

3 tablespoons brown sugar

2 teaspoons chipotle chili powder

2 teaspoons paprika

Thoroughly mix all the ingredients in a bowl and rub on the chicken breasts, legs, or wings before sealing in a bag. Store the rub in an airtight container, and use within 2 to 3 weeks before the sugar begins to clump.

PORK TENDERLOIN DRY RUB

Prep time: 5 MINUTES

Pork tenderloin is one of the best kept sous vide secrets. This tender cut of pork remains full of moisture and can be cooked to a more medium-rare finish compared with stovetop and oven-cooked tenderloin. This five-ingredient rub complements natural pork flavors and is very simple to whip up. ⅓ **CUP**

1 tablespoon ground cumin
1 tablespoon garlic powder
1 tablespoon dried rosemary

1 tablespoon dried thyme
Pinch lime sea salt

Thoroughly mix all the ingredients in a bowl and sprinkle all over the tenderloin, pressing down gently so the rub binds to the pork before sealing the pork in a bag. Store the rub in an airtight container, and use within 2 to 3 weeks.

Tip: If you can't find lime sea salt, use plain sea salt, and squeeze a few drops of lime juice into the salt or add a few strands of lime zest to the bag before sealing—you are looking for a faint lime flavor, not an overpowering one.

SALMON DRY RUB

Prep time: 5 MINUTES

Sous vide salmon can be done two different ways—dry rubbed, or poached in a marinade. I love both options, and they each offer different flavors. This rub has some sweetness and some spice and when combined, truly complements this popular fish. **¼ CUP**

2 tablespoons light brown sugar
½ tablespoon chili powder
½ tablespoon ground cumin
½ tablespoon paprika

1 teaspoon freshly ground black pepper
½ teaspoon dry mustard
½ teaspoon sea salt
¼ teaspoon ground cinnamon

Thoroughly mix all the ingredients in a bowl, making sure any sugar lumps are crushed, and rub onto both sides of the salmon before sealing in a bag. Store the rub in an airtight container, and use within 2 to 3 weeks.

BLACK BEAN STEAK SAUCE

Prep time: 20 MINUTES

Sous vide steak is delicious with just salt and pepper, but sometimes you want to dress things up a little. This homemade black bean sauce combines the earthiness of black beans with a little sweetness and umami from soy sauce and is delicious with flank steak and steak tacos. Use this as a sauce after cooking the steak as opposed to using as a marinade, otherwise you'll overpower the steak flavor. **1 CUP**

½ (14-ounce) can black beans
5 tablespoons water
1 red Anaheim chile pepper
2 tablespoons apple cider vinegar
2 teaspoons honey or raw sugar
2 teaspoons soy sauce

2 teaspoons tahini paste
1 teaspoon brown sugar
½ teaspoon crushed anise seeds
½ teaspoon crushed fennel seeds
½ teaspoon ground nutmeg
½ teaspoon grated fresh ginger

1. Drain and rinse the beans.

2. Add the beans, water, chile, cider vinegar, honey, soy sauce, tahini, brown sugar, anise, fennel, nutmeg, and ginger to a food processor and blend on high until smooth, about 5 minutes.

3. Pour the sauce into a small saucepan over medium heat for 5 minutes until simmering. Stir the sauce constantly until thickened, another 5 to 10 minutes.

4. Remove from the heat and serve with beef. Refrigerate any remaining sauce and use within 1 to 2 days.

TERIYAKI SAUCE

Prep time: 20 MINUTES | Cook time: 1 MINUTE

A traditional Japanese sauce, teriyaki sauce's flavor profile works well with many dishes. Chicken teriyaki, one of the most common teriyaki sauce dishes, is a tried-and-true favorite. Factor in sous vide cooking methods and the umami flavors, and you'll be able to produce restaurant-quality teriyaki dishes every single time. **¼ CUP**

¼ cup soy sauce

3 tablespoons white sugar

2 tablespoons mirin

1 garlic clove, minced

¼ teaspoon finely grated fresh ginger

¼ teaspoon cornstarch

Mix the soy sauce, sugar, mirin, garlic, ginger, and cornstarch together in a small saucepan over medium heat. When it begins to simmer, continue stirring until the sugar has dissolved, about 1 minute. Cool to room temperature before adding to the bag or your dish. Cover any remaining sauce, store in the refrigerator, and use within 1 to 2 days.

TEQUILA-LIME SAUCE

Prep time: 5 MINUTES | Cook time: 15 MINUTES

Most popular with chicken breasts and thighs, this fresh citrus sauce is a great summertime treat—think of it as a margarita for your chicken dishes. This is more of an adults-only option since we don't boil off all the alcohol during cooking. **1½ CUPS**

½ cup white tequila
2 tablespoons freshly squeezed lime juice
2 tablespoons minced shallot
1 tablespoon chopped cilantro
1 teaspoon pressed garlic

½ teaspoon salt
¼ teaspoon freshly ground black pepper
¼ cup heavy (whipping) cream
½ cup (8 tablespoons) unsalted
   butter, cubed

1. Combine the tequila, lime juice, shallot, cilantro, garlic, salt, and pepper in a medium saucepan over medium-high heat and bring to a boil. Boil until reduced by half, about 10 minutes.

2. Add the cream and stir often for 4 minutes.

3. Using a whisk, mix the butter into the sauce until fully melted.

4. Remove from the heat and serve as a dipping sauce or spoon over chicken breasts and legs.

PIRI PIRI SAUCE

Prep time: 20 MINUTES | Cook time: 30 MINUTES

My parents were born and raised in South Africa, so piri piri sauce was a mainstay in our house. Even though its roots are Portuguese, piri piri sauce continues to be a very popular flavor for braais (Afrikaans for barbecue or roast). You can add this sauce to a variety of dishes but the most common are roasted chicken and barbecued shrimp. The recipe is moderately spicy, but you can control the heat by reducing the pepper count to match your needs. Smaaklike ete! (Bon appétit!) **2½ CUPS**

2 large red bell peppers, halved and seeded
1 large red onion, cut into rings
6 red piri piri or bird's eye chiles
4 garlic cloves, minced
2 teaspoons kosher salt
1 teaspoon freshly ground black pepper
1 teaspoon dried oregano

¾ cup freshly squeezed lemon juice, divided
½ cup apple cider vinegar, divided
1½ teaspoons smoked paprika
Zest of 2 lemons, divided
2 bay leaves
½ cup extra-virgin olive oil

1. Heat a cast iron pan over medium-high heat, then add the bell peppers and onion rings and roast until charred, 4 or 5 minutes. Transfer to a cutting board and chop the peppers and onions into 2-inch pieces.

2. Purée the peppers and onions in a food processor until smooth. Add the chiles, garlic, salt, pepper, oregano, ½ cup of lemon juice, ¼ cup of cider vinegar, paprika, and zest of 1 lemon to the processor and purée until smooth.

3. Pour the sauce into a medium saucepan, place the bay leaves on top, and simmer over medium heat for 30 minutes. Remove and discard the bay leaves.

continued

4. Let the sauce cool, then transfer to the food processor and purée for 2 or 3 minutes. Add the remaining zest of 1 lemon, ¼ cup of lemon juice, and ¼ cup of cider vinegar and purée for 2 minutes.

5. Slowly add the olive oil to the running food processor and purée until well mixed.

6. Let cool to room temperature and use as a dipping sauce or marinade for chicken and shrimp dishes. Store any remaining sauce in the refrigerator in a canning jar or bottle for up to 30 days.

BUTTER CHICKEN SAUCE

Prep time: 5 MINUTES | Cook time: 35 MINUTES

Probably the most recognized Indian dish, butter chicken is surprisingly easy to make sous vide. You'll need to venture into the ethnic section of your local grocery store to find a few of the ingredients, but you'll be glad you did! This is a rich and creamy way to enjoy chicken and is a great entry point to Indian cuisine. **5 CUPS**

1 (14-ounce) can tomato sauce
3 cups heavy (whipping) cream
1 cup (2 sticks) butter, cubed, divided
2 tablespoons tandoori masala
2 teaspoons salt

1 teaspoon garam masala
1 teaspoon cayenne pepper
1 white onion, chopped
1 tablespoon finely chopped garlic

1. Add the tomato sauce, cream, ¾ cup of butter, the tandoori masala, salt, garam masala, and cayenne to a medium saucepan over medium heat and simmer for 30 minutes.

2. Meanwhile, melt the remaining ¼ cup of butter cubes in a separate small saucepan over medium heat. Add the onion and garlic and cook, stirring occasionally, for 15 minutes.

3. Add the onion mixture to the tomato-cream sauce and simmer for 5 minutes more.

4. Let cool for 10 to 15 minutes, stirring occasionally, before pouring the contents over the chicken. Seal before cooking.

FLANK, HANGER, FLATIRON, *and* SKIRT STEAK MARINADE

Prep time: 5 MINUTES

While these four cuts come from different parts of the cow, they are similar in that they are thin and tougher than the more tender beef cuts you can find. Because of their thinness, achieving rare or medium rare via traditional cooking can be difficult, but cooking them sous vide eliminates the issue. Combining precision cooking with a flavorful marinade means you get to enjoy the versatility of these cuts. **2 CUPS**

1 cup soy sauce
½ cup brown sugar
⅓ cup extra-virgin olive oil

4 garlic cloves, chopped
1 tablespoon sea salt
1 tablespoon freshly ground black pepper

1. Thoroughly mix all the ingredients together in a bowl.

2. Pour the marinade over the steak in a cooking bag, but reserve about ¼ cup for brushing on the beef before searing.

SWEET CHILI MARINADE

Prep time: 5 MINUTES | Cook time: 3 MINUTES

One of the biggest benefits of sous vide cooking, especially with chicken, is the ability to prepare the marinade, add the chicken (thighs or breasts), seal, and store in the refrigerator for a few hours to allow the chicken to soak up the flavors before you cook. Once you've reached the desired marinating time, simply warm up the water bath and transfer the marinated chicken directly to the water bath. This flavor is a personal favorite as I prefer things on the sweeter side, but I also enjoy a touch of heat. It's versatile and works well with chicken and rice, chicken salads, as a sauce for sweeter chicken wings, or even on sliced chicken breasts as a snack. **1½ CUPS**

½ cup sweet chili sauce

3 tablespoons soy sauce

2 tablespoons freshly grated ginger

2 garlic cloves, minced

1 tablespoon honey or raw sugar

Juice of 1 lime

1. Combine all the ingredients in a small saucepan over medium-high heat and cook for 2 to 3 minutes.

2. Let cool for 10 to 15 minutes, stirring occasionally, before pouring over chicken breasts, legs, or thighs in a bag and sealing.

MAUI PORK MARINADE

Prep time: 10 MINUTES

When I was a boy, my family used to vacation in Maui every year. I have very fond memories of the warm oceans with big surf, soft and long sandy beaches, and of course, the food. I remember eating Maui ribs for the first time, and watching the chef dig out the ribs from under the sand where they had cooked on hot lava rocks for hours on end. While we can't all wrangle up some hot lava rocks and a sandy beach, we can capture some Hawaiian flavors in our pork dishes. This marinade works well with pork chops, tenderloin, or ribs. **3 CUPS**

2 cups pulsed pineapple or unsweetened pineapple juice (with pulp)
4 tablespoons minced fresh ginger
4 tablespoons soy sauce
3 tablespoons sesame oil

6 shallots, finely chopped
3 tablespoons brown sugar
¾ teaspoon freshly ground black pepper
½ teaspoon sea salt

1. Thoroughly mix all the ingredients together in a bowl.

2. Add to the cooking bag with the pork and seal. Marinate for 4 to 6 hours before adding to a water bath.

BALSAMIC *and* MINT MARINADE

Prep time: 10 MINUTES

A lamb dish is an Easter favorite for many households, and no lamb dish is complete without some mint. While most racks or legs of lamb are traditionally served with mint sauce (or jelly) on the side, with this recipe you can capture that flavor in the marinade itself. While not as powerful as a mint sauce or jelly, the combination of balsamic and mint points this lamb dish in a subtler but equally delicious direction. ½ CUP

¼ cup fresh mint leaves, chopped
1 tablespoon extra-virgin olive oil
1 tablespoon balsamic vinegar
2 teaspoons honey or raw sugar

2 teaspoons ground cumin
1 teaspoon dry mustard
½ teaspoon salt
½ teaspoon freshly ground black pepper

1. Thoroughly mix all the ingredients together in a bowl.

2. Add to the cooking bag with the lamb and seal. Marinate for 4 hours before adding to a water bath.

PEANUT CHICKEN MARINADE

Prep time: 10 MINUTES

A can't-miss favorite, this creamy marinade can double as a sauce and infuses chicken with traditional Thai flavors while keeping the meat moist and full of nutty goodness. Peanut chicken can be enjoyed in many ways, so this recipe can be used over and over. Try this sauce as a salad dressing, dipping sauce for chicken satay, or as a pasta sauce with fresh noodles. ½ **CUP**

1 cup creamy peanut butter

⅔ cup coconut milk

⅓ cup soy sauce

¼ cup rice vinegar

3 tablespoons honey or raw sugar

2 garlic cloves, minced

1½ teaspoons minced fresh ginger

¼ teaspoon red pepper flakes

1. Whisk together all the ingredients in a bowl until well mixed.

2. If the marinade is too thick, add 1 tablespoon of water and mix until smooth.

3. Add to the cooking bag with the chicken and seal. Marinate for 2 to 4 hours before adding to a water bath.

POACHED EGGS TWO WAYS

Cook time: 13 MINUTES | Finishing time: 1 MINUTE | **167°F (75°C)**
Cook time: 45 MINUTES | Finishing time: 1 MINUTE | **145°F (62.8°C)**

In our office, there is an ongoing debate about egg texture. I prefer a 13-minute egg; I like my whites to be a bit more viscous and the yolk to be runny when I open the egg with a fork, and I like a quicker breakfast. The other side of the aisle prefers the 45-minute egg with firmer whites and a thicker but still soft and runny yolk. They use that extra time in the morning to prep for work. My best advice here is to try both and see which method produces the poached egg you prefer.

1 or more eggs (you can cook a dozen or more if you have the capacity in your container)

1 tray ice cubes

1. Set the sous vide circulator to 167°F (75°C) for runnier white and yolk or 145°F (62.8°C) for firmer white, thicker but still soft yolk.

2. Using a slotted spoon, gently lower the egg(s) directly into the water bath and let them rest on the bottom of the container.

3. Cook the eggs for the desired length of time (13 minutes at 167°F [75°C] or 45 minutes at 145°F [62.8°C]).

4. 5 minutes before the eggs are cooked, prepare an ice bath: Add the tray of ice cubes to a bowl and add water.

5. Transfer the cooked eggs to the ice bath and submerge for 1 minute (this is to help with handling the hot shells).

6. Carefully crack the shell and twist the upper portion to remove.

7. Before plating each egg, carefully transfer it to a slotted spoon to drain off any watery egg white.

HARD-BOILED EGGS

Cook time: 1 HOUR | 167°F (75°C)

This is a very simple method for preparing hard-boiled eggs that gives you a predictable outcome each time. If you're preparing lunches or snacks for the long workweek ahead, this is an effective way to produce lots of eggs for the family with very little effort or monitoring.

1 or more eggs (you can cook a dozen or more if you have the capacity in your container)

1. Set the sous vide circulator to 167°F (75°C).

2. Gently lower the egg(s) into the water and cook for 60 minutes.

3. Remove the eggs and let cool before peeling.

Tip: Peeling hard-boiled eggs can be an exercise in patience, and if you're planning on cooking more than a few of these in one batch, a shortcut might be welcome. If you boil the eggs for 3 minutes before placing them in the water bath for the sous vide portion, you'll find it much easier to remove the shells once they are cooked.

SOFT-BOILED EGGS

Cook time: 1 HOUR | 143°F (61.6°C)

Growing up, soft-boiled "eggs and soldiers" were one of my favorite meals. Fast-forward 30 years and I see an opportunity to rekindle this memory. While you're using your stove top for dinner, pop an egg or two into the sous vide circulator and forget it. One hour later, pop some bread in the toaster, and you've got a reliable egg dish for little ones that you can time to match your other dinner dishes so everyone can enjoy dinner together.

1 or more eggs (you can cook a dozen or more if you have the capacity in your container)

1. Set the sous vide circulator to 143°F (61.6°C).

2. Gently lower the egg(s) into the water and cook for 60 minutes.

3. Remove the eggs and serve or store in the refrigerator for 2 to 3 days.

SCRAMBLED EGGS

Prep time: 5 MINUTES | Cook time: 40 MINUTES | Finishing time: 1 MINUTE | 170°F (76.7°C)

While not the most efficient way to prepare scrambled eggs, it certainly will produce the richest, creamiest custard-like eggs you'll ever have. Heston Blumenthal once produced these eggs for another sous vide master, Kenji López-Alt, to demonstrate the capabilities of sous vide cooking. Kenji said they were the best eggs he'd ever tasted. **4 SERVINGS**

8 large eggs
½ cup grated medium Cheddar cheese
4 tablespoons heavy (whipping) cream
4 tablespoons 2% milk

Pinch salt
2 tablespoons butter
Finely chopped fresh chives (optional)

1. Set the sous vide circulator to 170°F (76.7°C).

2. Whisk together the eggs, cheese, cream, milk, and salt until well mixed.

3. Pour into a vacuum bag or large food-safe resealable bag. Add the butter. If using two smaller bags, halve the ingredients and mix in separate bowls to ensure equal distribution. If vacuum sealing the contents, move the sealer closer to the edge of the counter and drape the bag over the edge. This prevents the vacuum sealer from sucking up the egg mix into the sealer itself.

4. Place the bag in the water bath, ensuring it is completely submerged, and cook for 20 minutes.

5. Remove the bag, lay it flat on a smooth surface, and massage the egg mixture to ensure even cooking. Note: *The bag and contents will be hot. Use oven mitts when massaging the bag.*

6. Return the bag to the water for an additional 20 minutes.

7. Remove the eggs and pour into a serving dish. Top with the chives (if using) and serve.

Tip: Depending on the richness desired, you can replace the heavy cream with milk (skim or 2%), which would produce lighter, less creamy/heavy scrambled eggs.

COFFEE SHOP EGG BITES

Cook time: 1 HOUR | **173°F (78.3°C)**

The omelet station is a brunch favorite because you can choose add-ins from available ingredients, and within a few minutes you're enjoying an omelet made just for you. The same goes for this recipe—add whatever you like. These little jars of goodness, made very popular recently by a large coffee chain, are great for breakfast, lunch, or dinner and can be made in bulk. I happen to love bacon and cheese, so that's the basis for this recipe, but I encourage you to add your own favorites. **6 SERVINGS**

6 eggs
½ cup grated applewood-smoked
   Cheddar cheese
¼ teaspoon salt

⅛ teaspoon freshly ground black pepper
3 slices cooked bacon cut into ½-inch
   pieces

1.   Set the sous vide circulator to 173°F (78.3°C).

2.   Combine the eggs, cheese, salt, and pepper in a food processor and mix until smooth.

3.   Pour the egg mixture evenly among six 6-ounce mason jars.

4.   Evenly distribute the chopped bacon over the egg mixture and stir gently to mix.

5.   Screw on the jar lids to fingertip tightness (see tip—overtightening will cause the jars to break).

6.   Place the jars in the water bath for 1 hour. Enjoy the egg bites right from the jar or store them in the refrigerator for up to 1 week.

Tip: When cooking with mason jars, screwing the lids on "to fingertip tightness" means twisting the lids until they feel firmly closed but could still be opened with your fingertips. This will ensure that there is some room for air to escape from the jars, which relieves the pressure that might otherwise cause the jars to crack.

chapter three
Vegetables

~~~~~~

One of the biggest benefits of cooking sous vide is that the nutrients often lost using traditional cooking methods are completely retained when cooking under vacuum. If we are going to eat our vegetables, we want to get the most out of them! You'll notice that the majority of the recipes in this chapter cook at the same temperature—185°F (85°C). This means that with some planning you can make a variety of different vegetable dishes with differing cook times using just one sous vide circulator. Not all vegetables cook at the same pace, so start with the ones that need the longest time, adding the ones with shorter cook times later. Your greens will be greener, your carrots a brighter orange, and the flavors you add to the bag permeate the veggies so well, you'll wonder how you managed to ever enjoy vegetables before cooking sous vide!

GARLIC-PARMESAN ASPARAGUS (page 57)

SPICY GARLIC GREEN BEANS

Prep time: 5 MINUTES | Cook time: 1 HOUR | Finishing time: 1 MINUTE | **186°F (85.6°C)**

One of my first memories of cooking was when my mom asked me to help remove the tops and tails of green beans. I thought it was such a treat. Maybe it was because I was helping or because she trusted me with a small knife, but either way, I enjoyed the task more back then than I do now. Green beans have great flavor on their own, even raw, but they also absorb flavors as they soften up while remaining crisp. Their natural sweetness is begging for a little heat to match the bright colors on the plate. **4 SERVINGS**

4 large handfuls fresh green beans, trimmed
1 teaspoon sea salt
1 tablespoon pressed garlic

1 tablespoon red pepper flakes
1 tablespoon sriracha sauce
1 tablespoon extra-virgin olive oil

1. Set the sous vide circulator to 186°F (85.6°C).

2. Bring a large saucepan of water to a boil over high heat. Blanch the beans for 60 seconds. Drain then immediately shock in a bowl of ice water.

3. Add the beans to a food-safe bag and sprinkle with the salt.

4. Mix together the garlic, red pepper flakes, sriracha, and olive oil in a small bowl.

5. Pour the sauce over the beans in a food-safe bag and seal. Place the sealed bag in the water bath and cook for 1 hour.

6. Remove the bag from the water bath and serve.

Tip: For even more flavor, after cooking, save the cooking liquid and pour it into a small saucepan. Cook over medium heat until reduced by half and thickened. Drizzle over the beans.

MINT GREEN PEA PURÉE

Prep time: 5 MINUTES | Cook time: 45 MINUTES | Finishing time: 5 MINUTES | **185°F (85°C)**

I ate green peas often growing up since they were my father's favorite. I later took a thirty-year hiatus from them, but when I started cooking sous vide, I also subscribed to the "sous vide all the things" mentality. After having success with cauliflower purée, I knew I'd found a way to reintroduce my old vegetable nemesis. The dish is easy to make, looks bright and fresh on the plate, and with a little creativity, can add a real restaurant look to your dish. **6 SIDE SERVINGS**

1 pound fresh green peas
Zest of 1 lemon
2 tablespoons finely chopped fresh mint
1 tablespoon extra-virgin olive oil

1 teaspoon freshly ground black pepper
1 teaspoon sea salt
¾ teaspoon garlic powder
½ cup vegetable broth (optional)

1. Set the sous vide circulator to 185°F (85°C).

2. Combine the peas, lemon zest, mint, olive oil, pepper, salt, and garlic powder in a food-safe bag and seal using your vacuum sealer or the water displacement method. Place the sealed bag in the water bath and cook for 45 minutes.

3. Transfer the cooked peas and cooking liquid to a food processor or blender and purée the peas.

4. To thin the purée, add half of the vegetable broth (if using) and purée again. If necessary, add more broth until your desired consistency is reached.

5. Spoon the purée either directly over chicken or fish, or spread on a plate using the underside of a spoon.

CRISPY-SKIN SMASHED POTATOES

Prep time: 5 MINUTES | Cook time: 1 HOUR, 15 MINUTES | Finishing time: 15 MINUTES | 190°F (87.8°C)

This hearty side dish is perfect for any season, and the combination of crispy and soft is sure to please any crowd. The trick to the crispy skin is roasting the potatoes after the sous vide cook. That, combined with the garlic infusion that occurs during the sous vide process, adds deep flavors and textures to basic roasted potatoes. **4 TO 6 SERVINGS**

2 pounds russet or Yukon gold potatoes, unpeeled

10 garlic cloves, chopped

4 teaspoons sea salt

6 tablespoons extra-virgin olive oil, divided

5 or 6 fresh thyme sprigs, divided

¼ cup duck, chicken, or bacon fat

1. Set the sous vide circulator to 190°F (87.8°C).

2. Combine the potatoes, garlic, salt, olive oil, and 2 or 3 thyme sprigs in a food-safe bag and seal using your vacuum sealer or the water displacement method. Place the sealed bag in the water bath and cook for 1 hour, 15 minutes.

3. Preheat the oven to 500°F (260°C).

4. Remove the potatoes from the water bath and transfer them to a large mixing bowl. Pour the cooking liquid into a small bowl and set aside.

5. Add the duck fat and remaining 2 or 3 thyme sprigs and toss the potatoes until they are well coated with fat.

6. Transfer to an aluminum foil–lined baking sheet and use the base of a glass to split the potatoes open. Drizzle with the reserved cooking liquid.

7. Roast the potatoes for 10 minutes, flipping them after 5 minutes.

8. Transfer to a serving dish, discard the thyme sprigs, and serve.

SWEET POTATO FRIES *with* MALT VINEGAR MAYONNAISE

Prep time: 10 MINUTES | Cook time: 1 HOUR, 30 MINUTES | Finishing time: 10 MINUTES | **185°F (85°C)**

This healthier alternative to traditional potato fries is worth the extra effort. These bite-size snacks are great as a side with chicken or steak, or just on their own. They also add some often-needed color and pop on the plate and are packed with nutritious sweetness and a touch of salty goodness. **4 TO 6 SERVINGS**

FOR THE FRIES

2 large sweet potatoes, peeled and cut into 3- to 4-inch-long, 1-inch-wide strips
Pinch plus ¼ teaspoon kosher salt, divided
1½ tablespoons vegetable oil
1 teaspoon smoked paprika
½ teaspoon cayenne pepper
¼ teaspoon freshly ground black pepper

FOR THE MALT VINEGAR MAYONNAISE

1 egg yolk
2 tablespoons malt vinegar, divided
¾ cup vegetable oil
Kosher salt
Freshly ground black pepper

TO MAKE THE FRIES

1. Set the sous vide circulator to 185°F (85°C).

2. Add the sweet potatoes and pinch of salt to a food-safe bag, making sure the fries are in a single layer to ensure even cooking, and seal the bag using your vacuum sealer or the water displacement method. Place the sealed bag in the water bath and cook for 1 hour, 30 minutes.

3. Remove the sweet potatoes from the bag and discard any cooking liquid. Pat the fries dry.

4. Heat a skillet over medium-high heat and add the vegetable oil. After 3 or 4 minutes, add the fries to the hot oil and sprinkle with the paprika, cayenne, the remaining ¼ teaspoon of salt, and pepper.

5. Shake the potatoes and mixed spices into a single layer and cook the fries until golden brown on all sides, 5 to 7 minutes.

1. Whisk together the egg yolk and 1 tablespoon of vinegar. Very slowly pour in the vegetable oil while continuing to briskly whisk until all the oil has been added and the mayonnaise is thick and smooth.

2. Add the remaining 1 tablespoon of vinegar and season with salt and pepper. Stir to mix and serve with the fries.

TOMATO SOUP *with* GRILLED CHEESE SOLDIERS

Prep time: 10 MINUTES, PLUS 30 MINUTES DRAINING TIME | Cook time: 40 MINUTES
Finishing time: 5 MINUTES | 185°F (85°C)

Great for a cold winter's night, this timeless classic can be made from scratch instead of the can, giving you more control over your flavors and the knowledge that you have used entirely fresh ingredients. I remember the sense of accomplishment I felt the first time I made soup from scratch. Here, the reward is rich and flavorful and lasts for days (literally!). **6 TO 8 SERVINGS**

FOR THE SOUP

2 pounds ripe heirloom tomatoes, cored
   and cut into 1- to 2-inch wedges
2 teaspoons kosher salt, divided
2 tablespoons extra-virgin olive oil
2 garlic cloves, minced
½ teaspoon chopped fresh thyme
⅛ teaspoon freshly ground black pepper
⅛ teaspoon red pepper flakes
½ teaspoon sherry vinegar
½ cup mascarpone cheese

FOR THE GRILLED CHEESE SOLDIERS

2 tablespoons butter or margarine
4 slices white or whole-wheat bread
2 slices Cheddar cheese

TO MAKE THE SOUP

1.   Set the sous vide circulator to 185°F (85°C).

2.   Place the tomatoes in a strainer over a large bowl. Sprinkle with 1 teaspoon of salt and toss to mix. Let the tomatoes rest in the strainer for 30 minutes. Discard the juices and water from the bowl.

3.   Combine the tomatoes, the remaining 1 teaspoon of salt, olive oil, garlic, thyme, pepper, and red pepper flakes in a large food-safe bag and seal using your vacuum sealer or water displacement method. Place the sealed bag in the water bath and cook for 40 minutes.

4. Pour the entire contents of the bag into a blender and add the sherry vinegar and mascarpone. Blend until you reach your preferred consistency, anywhere between slightly chunky and very smooth.

5. Taste the soup and season with additional salt and pepper as needed. Serve immediately or transfer to an airtight container to enjoy up to 5 days later.

TO MAKE THE GRILLED CHEESE SOLDIERS

1. Spread the butter over the bread slices, then place 2 slices of bread butter-side down in a skillet, top each bread slice with a slice of cheese, then top with the remaining 2 slices bread, butter-side up.

2. Cook over medium-high heat for 1 to 2 minutes before flipping. Repeat.

3. Remove from the heat, and slice into 2-inch-wide strips (4 or 5 long soldiers). Dip into the warm soup, and enjoy!

Tip: If you want a grilled cheese crouton that you add directly into the soup, cut the cheese sandwich into 1-inch cubes instead, then return them to the hot skillet for an additional minute or two to crisp them up further before adding to the soup.

HONEY-ROASTED CASHEW *and* BEET SALAD

Prep time: 15 MINUTES | Cook time: 1 HOUR, 15 MINUTES | Finishing time: 5 MINUTES | **185°F (85°C)**

Salads with sous vide elements aren't something you come across every day, but one thing we do know is that sous vide cooking is great for tenderizing vegetables while still maintaining their shape and form. Beets are an excellent example. When you combine their natural sweetness and nutty flavors with fresh greens, you'll have a summer salad that complements almost any meal. **6 TO 8 SERVINGS**

6 large beets, peeled and cut into 1-inch chunks

2 teaspoons kosher salt

2 teaspoons freshly ground black pepper

3 tablespoons honey or raw sugar, divided

2 tablespoons butter

Zest of 1 large orange

1 tablespoon extra-virgin olive oil

1 teaspoon sea salt

½ teaspoon cayenne pepper

1½ cups unsalted cashews

6 to 8 handfuls frisée lettuce

3 mandarin oranges, peeled and segmented

1 cup feta cheese, crumbled

1. Set the sous vide circulator to 185°F (85°C).

2. Place the beet chunks in a single layer in a large food-safe bag, and sprinkle with the salt and pepper.

3. Add 2 tablespoons of honey, the butter, and orange zest to the bag and seal using your vacuum sealer or the water displacement method. Place the sealed bag in the water bath and cook for 1 hour, 15 minutes.

4. Meanwhile, preheat the oven to 350°F (180°C).

5. Combine the remaining 1 tablespoon of honey, olive oil, salt, and cayenne in a medium mixing bowl and stir. Add the cashews to the bowl and toss to coat thoroughly.

6. Spread the cashew mix evenly on a wax paper–lined baking tray and bake for 10 minutes, stirring once after 5 minutes. Set aside and let cool.

7. When the beets are cooked, remove them from the pouch and discard the cooking liquid.

8. Place a handful of frisée on each plate, and top with orange segments and a spoonful of beets. Garnish with the feta and cashews and serve.

SOUS VIDE BUTTERED CORN *on* THE COB

Prep time: 5 MINUTES | Cook time: 30 MINUTES | Finishing time: 0 MINUTES | 185°F (85°C)

Corn that butters itself while it cooks? That's exactly what you get with this super simple recipe. It's quick and packed with flavor, and if you're having a large group for dinner, you can scale this up to match demand without breaking the bank or your back. My corn consumption has increased threefold since we first made this recipe. **6 SERVINGS**

6 ears yellow corn, husked
6 tablespoons salted or unsalted butter
Kosher salt

1. Set the sous vide circulator to 185°F (85°C).

2. Add up to 3 ears of corn to each bag depending on the size of your bags (you can cut the cobs in half to accommodate smaller bags).

3. Add 1 tablespoon of butter and 1 pinch of salt per ear of corn to the bag. Seal using your vacuum sealer or the water displacement method. Place the sealed bag in the water bath and cook for 30 minutes. Carefully remove the corn from the bag and serve.

Tip: As the corn heats up, the bag will start to fill with air and it will float to the top. The quickest and easiest way to counter this is to place a dinner plate in the water, on top of the bags of corn. It might not be elegant, but it works. Just check in occasionally to ensure the corn doesn't get past the plate and pop up to the top.

GARLIC *and* HERB CHEESE CAULIFLOWER PURÉE

Prep time: 5 MINUTES | Cook time: 1 HOUR, 30 MINUTES | Finishing time: 5 MINUTES | 185°F (85°C)

This recipe was originally created as an alternative to mashed potatoes. It was supposed to be a healthier option, but the thing about cauliflower is that it can be bland. My mom used to make homemade cheese sauce and I'd drown my dinner in it. I remembered that when I was looking for something to give the purée flavor and decided to try my favorite cheese, Boursin. The dish is rich and creamy and while not 100 percent healthier than potatoes, it's totally worth it. **6 SERVINGS**

1 large head white cauliflower, cut into 1- to 2-inch florets
3 tablespoons unsalted butter
1 teaspoon kosher salt

½ teaspoon freshly ground black pepper
5 ounces Boursin Garlic and Fine Herbs cheese

1. Set the sous vide circulator to 185°F (85°C).

2. Rinse the florets under cold water in a colander and shake to remove any remaining water.

3. Add the florets, butter, salt, and pepper to a large food-safe bag (or use two bags if necessary) and seal using your vacuum sealer or the water displacement method. Place the sealed bag in the water bath and cook for 1 hour, 30 minutes, using a dinner plate or other weight to keep the bag from floating to the surface.

4. Remove the bag(s) from the water and pour the entire contents into a blender or food processor.

5. Add the Boursin and blend on high for 3 to 4 minutes until smooth, and the mixture has the consistency of mashed potatoes.

6. Season with more salt and pepper if necessary. Serve immediately or chill for later use.

Tip: Boursin is available in different flavors. Try them for a variety of purée tastes!

GARLIC-PARMESAN ASPARAGUS

Prep time: 5 MINUTES | Cook time: 14 MINUTES | Finishing time: 2 MINUTES | 185°F (85°C)

Asparagus season is all too short so I do my best to make the most of it. Cooking this veggie sous vide results in a fresh, vibrant color that really adds a bright pop to your dinner plate. The sous vide method retains asparagus's moisture while still producing a firm stalk cooked to perfection. **4 TO 6 SERVINGS**

1 bunch green asparagus, trimmed

4 tablespoons unsalted butter, cut into cubes

Sea salt

1 tablespoon pressed garlic

¼ cup shaved Parmesan cheese

1. Set the sous vide circulator to 185°F (85°C).

2. Lay the asparagus in a single layer in a food-safe bag or use two bags if necessary.

3. Put the cubes of butter in the four corners of the bag and 1 tablespoon of garlic in the center. Add a few pinches of salt and seal the bag using your vacuum sealer or the water displacement method.

4. Massage the asparagus to distribute the garlic as evenly as possible then place the bag in the water bath and cook for 14 minutes.

5. Transfer the asparagus to a serving dish, drizzle with a bit of the cooking liquid, and plate.

6. Top with the Parmesan and serve immediately.

ASIAGO BROCCOLI *and* PEPPERS

Prep time: 10 MINUTES | Cook time: 40 MINUTES | Finishing time: 2 MINUTES | 185°F (85°C)

Although broccoli was never my favorite dish growing up, I have a better appreciation for this affordable, bright, and healthy vegetable that's a great side to almost any dish. As with cauliflower, I seem to need something to go over my broccoli—typically cheese. In addition, the red, yellow, and orange pepper strips add a contrasting pop of color. **4 TO 6 SERVINGS**

1 large head broccoli, cut into 1- to 2-inch florets (4 cups)

2 garlic cloves, minced, divided

1 teaspoon kosher salt, divided

1 teaspoon freshly ground black pepper, divided

2 tablespoons unsalted butter, divided

1 tablespoon extra-virgin olive oil, divided

½ large red bell pepper, cut into ¼-inch strips

½ large yellow bell pepper, cut into ¼-inch strips

½ large orange bell pepper, cut into ¼-inch strips

½ cup shaved Asiago cheese

1. Set the sous vide circulator to 185°F (85°C).

2. Add the broccoli florets, 1 clove of garlic, ½ teaspoon of salt, ½ teaspoon of pepper, 1 tablespoon of butter, and ½ tablespoon of olive oil to a medium bowl and toss.

3. In another medium bowl, add the bell peppers, the remaining 1 clove of garlic, remaining ½ teaspoon of salt, remaining ½ teaspoon of pepper, remaining 1 tablespoon of butter, and the remaining ½ tablespoon of olive oil and toss.

4. Transfer the broccoli and peppers to separate food-safe bags, and seal using your vacuum sealer or the water displacement method. Place the sealed bags in the water bath and cook for 40 minutes.

5. Using a slotted spoon, transfer the broccoli and bell peppers to a serving bowl, and discard the cooking liquids.

6. Toss the veggies, sprinkle with the shaved Asiago while the vegetables are still warm, and serve.

BUTTERNUT SQUASH SOUP

Prep time: 15 MINUTES | Cook time: 2 HOURS | Finishing time: 5 MINUTES | **185°F (85°C)**

A fall favorite in our household, butternut squash soup is a healthy, hearty, and heavenly dish that is surprisingly uncomplicated to make. As a mostly hands-off recipe, you'll be able to have fresh homemade soup ready to go while you work on other dishes for a multicourse meal (think Thanksgiving or over the holidays). The soup can also be made vegan (by omitting the cream and crème fraîche) if you need to accommodate dietary restrictions. **6 SERVINGS**

1 butternut squash, peeled and cut into 1- to 2-inch chunks (6 cups)

1 large Granny Smith apple, peeled, cored, and cut into 1- to 2-inch chunks, divided

½ yellow onion, finely diced

3 fresh thyme sprigs

1 garlic clove, finely chopped

1 teaspoon ground cumin

1 teaspoon sea salt

¾ cup heavy (whipping) cream or vegetable stock

½ teaspoon freshly ground black pepper

4 tablespoons crème fraîche

1. Set the sous vide circulator to 185°F (85°C).

2. Combine the squash, apple (reserving a few chunks for garnish), onion, thyme, garlic, cumin, and salt in a food-safe bag and seal using your vacuum sealer or the water displacement method. Place the sealed bag in the water bath and cook for 2 hours.

3. Transfer the contents of the bag, including any liquid, to a blender or food processor and blend until smooth. While blending, keep an eye on the consistency and slowly add the cream. Continue to blend until the desired consistency is reached.

4. Season with additional salt, if necessary, and the pepper.

5. Pour the soup into bowls and top each serving with some crème fraîche. Garnish with the reserved apple and serve.

MEATLESS STUFFED PEPPERS

Prep time: 20 MINUTES, PLUS 2 HOURS SOAKING TIME | Cook time: 1 HOUR, 15 MINUTES
Finishing time: 0 MINUTES | **185°F (85°C)**

A vegetarian play on classic stuffed-pepper recipes, this side dish is perfect for meat-averse guests. The "meat" is a mix of vegetables, spices, garlic, and nuts that creates a consistency, look, and taste that can replace ground beef. The added benefit of going this route is that you can cook all the ingredients at a much higher temperature—as required for vegetables—than you could cook beef. Shout out to Bekkah, a client and friend of mine, for inspiring this recipe!

4 TO 6 SERVINGS

4 carrots, chopped into small chunks

4 garlic cloves, minced

4 shallots, diced

1 cup raw cashews, soaked for 2 hours then drained

1 cup raw walnuts, soaked for 2 hours then drained

1 tablespoon balsamic vinegar

1 tablespoon tamari

1 tablespoon ground cumin

2 teaspoons paprika

1 teaspoon garlic powder

1 pinch cayenne pepper (or more if you want to add some heat)

4 fresh thyme sprigs

Zest of 1 lemon

4 bell peppers, tops cut off and saved, seeded

1. Set the sous vide circulator to 185°F (85°C).

2. Add the carrots, garlic, shallots, cashews, walnuts, balsamic vinegar, tamari, cumin, paprika, garlic powder, cayenne, thyme, and lemon zest to a food processor and blend until coarsely mixed (do not purée—you are trying to create the look of ground beef).

3. Spoon the veggie mixture into the bell pepper shells, pressing down to ensure a tight fit. Replace the top of each pepper.

4. Add 1 or 2 peppers to each food-safe bag and seal using your vacuum sealer or the water displacement method. Place the sealed bags in the water bath and cook for 1 hour, 15 minutes.

5. Remove the bell pepper tops and serve.

BUTTER POACHED HONEY-GLAZED CARROTS

Prep time: 5 MINUTES | Cook time: 1 HOUR | Finishing time: 3 MINUTES | 185°F (85°C)

The bright colors of sous vide vegetables are never more present than with sous vide carrots. They keep their shape, firmness, nutrients, and of course, their orange color. This is another one of my childhood favorites. My whole family had a serious love for carrots, including our dog Mandy. This is an easy recipe that should keep the whole family interested in clearing their plate of all things vegetable. **4 SERVINGS**

1 pound carrots, peeled and trimmed

3 tablespoons unsalted butter

2 tablespoons honey

1 teaspoon extra-virgin olive oil

1 teaspoon kosher salt

1 teaspoon freshly ground black pepper

1 tablespoon chopped fresh chervil

1. Set the sous vide circulator to 185°F (85°C).

2. Place the carrots, butter, honey, olive oil, salt, and pepper in a food-safe bag and seal the bag using your vacuum sealer or the water displacement method. Place the sealed bag in the water bath and cook for 1 hour.

3. Heat a cast iron pan over medium-high heat, pour the entire contents of the bag into the pan, and cook until the liquid begins to turn into a glaze, about 2 minutes.

4. Add the chervil to the pan and mix quickly before serving.

POTATO LEEK SOUP

Prep time: 10 MINUTES | Cook time: 2 HOURS | Finishing time: 5 MINUTES | **185°F (85°C)**

This fall and winter classic is a can't-miss soup for a snowy Sunday evening—smooth, creamy, and hearty enough to be served as a meal. Once it's cooked, it can be frozen for up to a week. The soup is often served with butter and bread or topped with crispy bacon, but garnishes and finishing touches are limitless with this versatile dish. **8 SERVINGS**

4 Yukon Gold potatoes, peeled and cut into 1-inch slices
1 yellow onion, cut into ¼-inch pieces
1 celery stalk, cut into ½-inch pieces
4 cups ½-inch diced leeks, white parts only
1 cup chicken or vegetable stock

8 tablespoons unsalted butter
4 garlic cloves, minced
2 bay leaves
1 teaspoon kosher salt
1 teaspoon freshly ground black pepper
2 cups heavy (whipping) cream
¼ cup chopped fresh chives

1.   Set the sous vide circulator to 185°F (85°C).

2.   Add the potatoes, onion, celery, leeks, stock, butter, garlic, and bay leaves to a food-safe bag and seal using your vacuum sealer or the water displacement method. Place the sealed bag in the water bath and cook for 2 hours.

3.   Discard the bay leaves and pour the entire contents of the bag into a food processor or blender. Add the salt and pepper and blend, while slowly pouring in the cream, until smooth, 2 to 3 minutes.

4.   Strain through a chinois or fine mesh strainer into a large serving bowl. Ladle into bowls, sprinkle with the chives, and serve.

CHICKEN CAESAR SALAD WRAPS

Prep time: 5 MINUTES | Cook time: 1 HOUR, 30 MINUTES | Finishing time: 5 MINUTES | 150°F (65.5°C)

Chicken breasts yield some of the best results that sous vide cooking delivers. It's also a great way to dress up your salads with protein. Chicken breasts served as a main dish typically are cooked at a lower temperature than chicken added to soups or salads. The tender and juicy results of cooking at 140°F aren't always ideal when chicken is used as a topping for other dishes. For that reason, we increase the temperature when chicken isn't the focus of a particular dish, like here. Increasing the temperature provides a more traditional texture that is closer to what you would find at a beachside lunch spot. **4 SERVINGS**

4 skinless, boneless chicken breasts
1 teaspoon kosher salt, divided
1 teaspoon freshly ground black pepper, divided
1 cup mayonnaise
2 garlic cloves, minced
2 tablespoons freshly squeezed lemon juice

1 teaspoon anchovy paste (optional)
1 teaspoon Dijon mustard
1 teaspoon Worcestershire sauce
4 cups loosely packed washed romaine lettuce
4 (10-inch) tortillas

1. Set the sous vide circulator to 150°F (65.5°C).

2. Season the chicken breasts with ¾ teaspoon of salt and ¾ teaspoon of pepper and seal in two separate food-safe bags using your vacuum sealer or the water displacement method. Place the sealed bags in the water bath and cook for 1 hour, 30 minutes.

3. While the chicken is cooking, in a medium bowl, whisk together the mayonnaise, garlic, lemon juice, anchovy paste (if using), mustard, Worcestershire sauce, and the remaining ¼ teaspoon of salt and ¼ teaspoon of pepper. Store the mixture in the refrigerator until the chicken is cooked.

4. When the chicken is cooked, remove it from the bag and pat dry. Slice the breasts into 1-inch strips.

5. Remove the dressing from the refrigerator and combine with the romaine lettuce in a medium bowl, tossing until well mixed.

6. Spoon one-quarter of the chicken mixture into each tortilla, leaving 2 inches clear all around. Fold in the sides, but not so much that they touch, and then bring the bottom flap up and continue to roll. Cut in half and serve.

HOMEMADE HUMMUS

Prep time: 5 MINUTES, PLUS 12 HOURS SOAKING TIME | Cook time: 3 HOURS
Finishing time: 30 MINUTES | 195°F (90.5°C)

This recipe takes some time and some patience but you'll end up with a concentrated flavor that beats any store-bought hummus or any canned chickpeas. It is also significantly less expensive to make if you have some time and plan ahead just a little bit. While we cover the basics, you can decide how you want to finish it—anything from pine nuts and lemon to roasted red pepper or even bacon and onion. **6 TO 8 SERVINGS**

1½ cups dried chickpeas
2 quarts water, divided
¼ cup freshly squeezed lemon juice
¼ cup tahini or sesame seed paste
2 garlic cloves, minced

2 tablespoons extra-virgin olive oil
½ teaspoon ground cumin
½ teaspoon kosher salt
1 teaspoon paprika

1. Put the chickpeas in a food-safe plastic container and soak in 1 quart of fresh water for 12 hours. Make sure the chickpeas are covered by water at all times.

2. Set the sous vide circulator to 195°F (90.5°C).

3. Drain the chickpeas and fill a food-safe bag with the remaining 1 quart of water. Add the chickpeas and seal using your vacuum sealer or the water displacement method. Place the sealed bag in the water bath and cook for 3 hours, then transfer the bag to an ice bath to chill.

4. Meanwhile, in a food processor, process the lemon juice and tahini for 90 seconds, scraping the mixture from the sides of the processor until you have a consistent texture. Add the garlic, olive oil, cumin, and salt to the processor and process until smooth, about 30 seconds, scraping the sides as you go.

5. Remove the chickpeas from the ice bath and drain. For a smoother hummus, peel off and discard the chickpea skins.

6. Add half the chickpeas to the processor and process for 90 seconds, scraping down the sides, then add the remaining chickpeas (reserving a handful for garnish) and continue to process. If your hummus mix is too thick, add 1 tablespoon of water at a time while processing until the desired consistency has been reached.

7. Transfer the mixture to a serving plate, sprinkle with the paprika, and garnish with the reserved chickpeas. Serve with warm pita bread.

LEMONY GARLIC KALE SALAD

Prep time: 5 MINUTES | Cook time: 8 MINUTES | Finishing time: 2 MINUTES | 195°F (90.5°C)

Still a very popular superfood, kale is a mainstay in grocery stores around the world. It's inexpensive, very good for your health, and has a variety of uses including salads, smoothies, or, in this case, as a healthy side for chicken and fish dishes. This recipe is one of the easiest ones out there, and you'll end up with tasty fresh greens on your plate. **4 TO 6 SERVINGS**

4 to 6 handfuls fresh kale, stemmed

6 tablespoons extra-virgin olive oil, divided

2 garlic cloves, crushed

4 tablespoons freshly squeezed lemon juice

½ teaspoon kosher salt

1. Set the sous vide circulator to 195°F (90.5°C).

2. Toss the kale in 2 tablespoons of olive oil.

3. Add the kale to a food-safe bag and seal using your vacuum sealer or the water displacement method. Place the sealed bag in the water bath and cook for 8 minutes.

4. Whisk the remaining 4 tablespoons of olive oil, garlic, lemon juice, and salt together.

5. Remove the kale from the bag, plate, drizzle with the dressing, and serve.

SWEET CORN SALSA

Prep time: 5 MINUTES | Cook time: 45 MINUTES | Finishing time: 5 MINUTES | 185°F (85°C)

Once you've tried the Sous Vide Buttered Corn on the Cob (page 55), you'll want to find other ways to use that juicy corn! This salsa is sweet and spicy and is a great complement to fish, chicken, or tacos, or it can be served on its own as a traditional dipping salsa for nachos and tortillas. Once you perfect it, tinker with it and make your own versions, adding other fresh ingredients to enhance this super simple summer dish. **6 TO 8 SERVINGS**

4 ears yellow corn, husked
⅓ cup extra-virgin olive oil
¼ teaspoon kosher salt
¼ teaspoon freshly ground black pepper
1 large tomato, chopped
3 tablespoons freshly squeezed lime juice

2 garlic cloves, minced
1 jalapeño or other fresh hot chile
  pepper, seeded
4 scallions, green parts only, chopped
½ bunch fresh cilantro leaves,
  finely chopped

1. Set the sous vide circulator to 185°F (85°C).

2. Lightly brush the corn with a little olive oil and sprinkle with the salt and pepper.

3. Add two ears of corn to each of two food-safe bags depending on the size of your bags (you can cut the ears in half to accommodate smaller bags) and seal using your vacuum sealer or the water displacement method. Place the sealed bags in the water bath and cook for 45 minutes.

4. Meanwhile, combine the tomato, remaining olive oil, lime juice, garlic, jalapeño, scallions, and cilantro in a medium bowl and mix together.

5. Preheat a grill or skillet to hot. Remove the corn from the bag and discard any liquid. Add the corn to the grill or skillet and cook until lightly charred, 2 to 3 minutes. Let cool.

6. Using a sharp knife, cut the kernels from the cob and add to the tomato mixture. Toss to mix.

7. Transfer to a serving dish or bowl and serve with fresh fish, add to a salad, or enjoy with tortilla chips.

ROSE'S CHINESE CHICKEN SALAD

Prep time: 10 MINUTES | Cook time: 1 HOUR, 30 MINUTES | Finishing time: 10 MINUTES | 150°F (65.5°C)

When I left home for university, my mother handed me a handwritten collection of recipes. All her best recipes were there, including a summer chicken salad with Chinese dressing and crispy wontons. Incorporating sous-vide-cooked chicken to enhance this already amazing salad makes sense. Thanks, Mom!

6 TO 8 SERVINGS

4 large skinless, boneless chicken breasts
1 teaspoon kosher salt
1 teaspoon freshly ground black pepper
¼ cup granulated sugar
¼ cup soy sauce
3 tablespoons smooth peanut butter
3 tablespoons sesame oil
2 tablespoons vegetable oil

4 teaspoons white vinegar
½ teaspoon cayenne pepper
1 large head iceberg lettuce, trimmed and torn
3 or 4 scallions, chopped
¼ cup slivered almonds, toasted
¼ cup sesame seeds, toasted
2 cups wonton strips

1. Set the sous vide circulator to 150°F (65.5°C).

2. Season the chicken breasts with the salt and pepper.

3. Add the chicken to a food-safe bag and seal using your vacuum sealer or the water displacement method. Place the sealed bag in the water bath and cook for 1 hour, 30 minutes.

4. Meanwhile, whisk together the sugar, soy sauce, peanut butter, sesame oil, vegetable oil, vinegar, and cayenne in a medium bowl until smooth and place in the refrigerator.

5. When the chicken breasts are cooked, remove them from the bag, discard the cooking liquid, and pat the breasts dry.

6. Using a sharp knife, cut the chicken into 1- to 2-inch pieces. Transfer to a large salad bowl with the lettuce, scallions, and almonds.

7. Pour the dressing over the salad, toss to mix, sprinkle with the sesame seeds and wonton strips, and serve immediately.

Tip: To keep your salad crisp, place an inverted saucer in the bottom of the salad bowl to allow excess dressing to drain under the saucer—you'll never have a limp salad again!

VEGETABLE HALLOUMI KEBABS

Prep time: 15 MINUTES | Cook time: 45 MINUTES | Finishing time: 5 MINUTES | **185°F (85°C)**

One of the biggest challenges with making meat kebabs, especially using sous vide methods, is that you need at least two sous vide machines to accommodate the differences between the temperature at which you would cook beef, chicken, or lamb and the zucchini, peppers, squash, and onions. A few of our friends here in London have some different dietary restrictions, and the majority of those restrictions involve types of meats. This is an easy recipe to create a version without meat to accommodate dietary needs. **6 TO 8 SERVINGS**

1 large yellow squash, peeled and cut into 1- to 1½-inch slices

1 large eggplant, peeled and cut into 1- to 1½-inch slices

1 large yellow bell pepper, cut into 2-inch wide chunks

1 large red bell pepper, cut into 2-inch wide chunks

1 large green bell pepper, cut into 2-inch-wide chunks

2 (8-ounce) packages halloumi cheese

¼ cup extra-virgin olive oil

1 teaspoon honey

Sea salt

Freshly ground black pepper

1.   Set the sous vide circulator to 185°F (85°C).

2.   Fill a large bowl with water and soak the skewers while the ingredients cook.

3.   Add the squash and eggplant to separate food-safe bags. Add all the bell pepper chunks to one large food-safe bag. Seal all the bags using your vacuum sealer or the water displacement method. Place the sealed bags in the water bath and cook for 45 minutes.

4.   When 10 minutes of cooking time remain, heat a large pan over medium-high heat.

5.   Drain the halloumi, pat dry using a paper towel, and cut into 1-inch cubes. Brush all sides of the cheese with the olive oil, then add it to the skillet and sear until a brown crust begins to form on all sides. Remove from the heat, drizzle with the honey, and cover.

6.   When the vegetables are cooked, discard all cooking liquid and place the veggies in a medium bowl. Season with salt and pepper.

7.   Thread the veggies and cheese onto the skewers, alternating the squash, eggplant, bell peppers, and halloumi, and serve.

GARLIC BABY BOK CHOY

Prep time: 10 MINUTES | Cook time: 30 MINUTES | Finishing time: 3 MINUTES | **185°F (85°C)**

This Chinese cabbage is healthy, easy to make, and goes well with many recipes in this book. Typically steamed, we can use sous vide techniques to push the flavors of garlic, chiles, soy, and sesame into this bright and colorful green vegetable. Quickly prepare bok choy to serve with Miso-Glazed Chilean Sea Bass (page 75), Teriyaki and Ginger–Glazed Salmon (page 77) or Korean Kalbi Beef Ribs (page 158). **6 TO 8 SERVINGS**

1½ pounds bok choy

2 garlic cloves, minced

2 tablespoons vegetable oil

1 tablespoon soy sauce

1 teaspoon freshly grated ginger

¼ teaspoon red pepper flakes

¼ teaspoon toasted sesame oil

1 tablespoon sesame seeds, for garnish

1. Set the sous vide circulator to 185°F (85°C).

2. Slice each bok choy vertically in half.

3. Combine the garlic, vegetable oil, soy sauce, ginger, and red pepper flakes in a small saucepan over medium heat and cook for 4 to 5 minutes. Remove from the heat and let cool.

4. Place the bok choy in a single layer in a food-safe bag, pour the garlic-oil mixture into the bag, and massage with your hands to ensure all the pieces are well coated.

5. Seal using your vacuum sealer or using the water displacement method. Place the bag in the water bath and cook for 30 minutes.

6. Remove the bok choy from the bag (reserving the cooking liquids), place on a paper towel–lined baking sheet, and pat dry.

7. Put the sesame oil in a medium bowl, add the bok choy, and toss gently.

8. Transfer the bok choy to a serving dish, drizzle with a small amount of the reserved cooking liquid, and garnish with the sesame seeds before serving.

chapter four
Fish and Seafood

Cooking fish can be difficult! Different types of fish need to be cooked at different temperatures, some are more delicate than others, but all of them can be overcooked very easily, which is what makes cooking fish using traditional methods a challenge. Cooking seafood and fish sous vide removes the guesswork and allows you to tailor the cook temperature and time to match the consistency you love.

CAJUN SHRIMP SKEWERS (page 84)

LEMON *and* GARLIC SHRIMP LINGUINE

Prep time: 20 MINUTES | Cook time: 30 MINUTES | Finishing time: 5 MINUTES | 137°F (58.3°C)

This dish is packed full of fresh flavors and is perfect for a warm summer's evening. Shrimp and citrus are a classic pairing, but cooking them sous vide elevates their harmony—the citrus flavors are absorbed deep into the shrimp, resulting in a perfectly balanced, excellent seafood pasta that will impress any guest. **4 SERVINGS**

6 tablespoons unsalted butter

2 cups baby spinach, chopped

½ cup Parmesan cheese, divided

2 garlic cloves, pressed

Zest of 1 lemon

1½ tablespoons freshly squeezed lemon juice

1 tablespoon minced fresh parsley

1 teaspoon plus ⅛ teaspoon kosher salt, divided

1 teaspoon freshly ground black pepper

1 teaspoon red pepper flakes

1½ pounds jumbo shrimp, peeled and deveined, tails left on

8 ounces linguine

1. Set your sous vide circulator to 137°F (58.3°C).

2. Combine the butter, spinach, ¼ cup of Parmesan, garlic, lemon zest and juice, parsley, 1 teaspoon of salt, pepper, and red pepper flakes in a medium saucepan over low heat. Cook until the butter has melted, about 5 minutes. Stir to mix well, then remove from the heat. Let cool, stirring occasionally.

3. Add the shrimp to a food-safe bag and pour the cheese mixture into the bag, massaging the shrimp with the sauce to coat. Flatten the shrimp into a single layer before sealing with your vacuum sealer or using the water displacement method. Massage the shrimp again into a single layer if necessary. Place the sealed bag in the water bath and cook for 30 minutes.

4. Shortly before the shrimp are cooked, bring a large pot of water and the remaining ⅛ teaspoon of salt to a boil over medium-high heat. Add the linguine to the pot and cook according to the package directions. Drain and return the linguine to the pot.

5. When the shrimp are cooked, pour the entire contents of the bag over the pasta. Heat over medium heat for 3 to 4 minutes to warm through.

6. Transfer to a serving platter, top with the remaining ¼ cup of Parmesan, and serve.

GARLIC BUTTERED LOBSTER TAILS

Prep time: 10 MINUTES | Cook time: 50 MINUTES | Finishing time: 10 MINUTES | **137°F (58.3°C)**

For a few years, my family's Christmas dinner changed from the traditional roasted turkey to surf and turf. I loved it, but the tradition only lasted a few years since not everyone loves seafood (looking at you, Dad). While it lasted, we always grilled the lobsters. Grilled lobster is excellent, until you've had it cooked sous vide. Lobster can dry out quickly on the grill, but when you cook it sous vide, you won't have that problem. Pair with Chateaubriand (page 137), for a surf and turf holiday dinner. **4 SERVINGS**

6 garlic cloves, minced, divided
Zest of ½ lemon
½ cup freshly squeezed lemon
   juice, divided

4 medium to large lobster tails,
   in the shell
4 tablespoons unsalted butter, divided

1.   Set your sous vide circulator to 137°F (58.3°C).

2.   Mix together half the garlic, the lemon zest, and ¼ cup of lemon juice in a medium bowl. Set aside.

3.   Bring a large pot of water to a boil. Place the lobster tails (still in their shells) into the water for 90 seconds. Remove the tails and place on a clean surface to cool.

4.   When cool enough to handle, squeeze the tails inward from the edges to crack down the center of each shell. Carefully remove the meat from each tail.

5.   Add half of the lobster meat, 2 tablespoons of butter, and half of the lemon-garlic mix to each of two food-safe bags and seal using your vacuum sealer or using the water displacement method. Place the sealed bags in the water bath and cook the tails for 50 minutes.

6.   Remove the tails from the cooking bags and discard the liquid.

7.   Melt the remaining 2 tablespoons of butter, remaining lemon-garlic mixture, and remaining ¼ cup of lemon juice in a small saucepan over medium-low heat.

8.   Distribute the lobster among 4 ramekins and place over individual butter warmers at the dinner table. Serve the lobster immediately with the lemon butter.

MISO-GLAZED CHILEAN SEA BASS

Prep time: 10 MINUTES | Cook time: 30 MINUTES | Finishing time: 10 MINUTES | 134°F (56.6C)

The last time I officially worked in a restaurant was at Barracuda's in Mississauga, Ontario. The restaurant was billed as a "seafood-inspired fine dining restaurant." The chef was very talented, and his was my first try of Chilean sea bass—it was love at first bite. The fish goes by another name, Patagonian toothfish, but I'm not sure the restaurant would have sold as many dishes if they used that moniker. **4 SERVINGS**

2 tablespoons unsalted butter

⅓ cup sake

⅓ cup red or yellow miso

¼ cup mirin

3 tablespoons brown sugar

2½ tablespoons low-sodium soy sauce

4 (6-ounce) fillets Chilean sea bass or black cod

1 tablespoon extra-virgin olive oil or butter

2 tablespoons chopped scallions

2 tablespoons chopped fresh basil

1.　Set your sous vide circulator to 134°F (56.6°C).

2.　Melt the butter in a small saucepan over medium-low heat. Add the sake, miso, mirin, brown sugar, and soy sauce to the melted butter and stir for 1 minute. Remove from the heat and let cool.

3.　Place two fish fillets in each of two food-safe bags and distribute the miso mixture evenly between the bags. Seal using your vacuum sealer or the water displacement method. Place the bags in the water bath and cook for 30 minutes.

4.　When the fish is cooked, lay the bags on a flat surface and, using a spatula or fish lifter, carefully remove the fillets and transfer to paper towels. Transfer the cooking juices to a small saucepan over medium-high heat and cook, stirring often, until reduced by half.

5.　While the juices are reducing, heat the olive oil in a cast iron pan over medium-high heat and place the fillets in the pan, skin-side down, pressing down every 30 seconds until the skin is crisp, 1½ to 2 minutes.

6.　Plate the fish, skin-side down, and spoon the miso glaze over the fish. Top with the scallions and basil before serving over rice or with vegetables.

GRILLED OCTOPUS

Prep time: 5 MINUTES | Cook time: 4 HOURS | Finishing time: 10 MINUTES | 180°F (82.2°C)

I've always enjoyed grilled octopus. When my wife and I go out to eat, I'll order it every chance I get. I have a few friends who either own, manage, or cook at restaurants and they've all told me how difficult it is to get octopus right—a minute or two too long or too short and the dish is ruined. The precision of a sous vide circulator makes cooking octopus a reality at home. **4 SERVINGS**

1 pound octopus tentacles
Kosher salt
Freshly ground black pepper
5 tablespoons extra-virgin olive oil, divided

2 tablespoons freshly squeezed lemon juice
1 tablespoon freshly grated lemon zest
1 tablespoon minced fresh parsley

1. Set your sous vide to 180°F (82.2°C).

2. Using a cleaver or sharp knife, select as many tentacles as you need (up to 8!) and cut them into lengths. They will reduce in size significantly once they hit the water so don't cut them too short. Season with salt and pepper.

3. Add 2 tablespoons of olive oil to a food-safe bag, then place the tentacles in the bag and seal it using a vacuum sealer or the water displacement technique. Place the sealed bag in the water bath and cook for 4 hours. (You will notice that after about 30 minutes, the color of the oil-octopus mix will start to turn a shade of pink/purple—this is to be expected.)

4. When the octopus is cooked, cut the corner of the sealed bag (or open just the corner of a resealable bag) and pour out and discard the cooking liquid. Remove the tentacles and pat them dry with paper towels.

5. Brush all sides with the remaining 3 tablespoons of olive oil.

6. Heat a large cast iron pan over medium-high heat and cook the tentacles, turning them to ensure all sides are cooked evenly, about 10 to 15 seconds per side. Remove them from the pan.

7. Mix together the lemon juice, lemon zest, and parsley in a small bowl.

8. Drizzle the octopus with the lemon dressing. Before plating, sear the tentacles with a propane torch to add a crispy outer layer to the skin. Don't overdo it!

TERIYAKI *and* GINGER–GLAZED SALMON

Prep time: 10 MINUTES, PLUS 1 HOUR MARINATING TIME | Cook time: 40 MINUTES
Finishing time: 3 MINUTES | **128°F (53.3°C)**

Sous vide salmon is sweet and succulent and in this recipe, receives a great infusion of flavors from the marinade, as well as a hint of smokiness from finishing in a hot pan. Say good-bye to traditional grilled, dry salmon; you'll never go back after trying it sous vide. **4 SERVINGS**

½ cup soy sauce

6 tablespoons white sugar

4 tablespoons mirin

2 small garlic cloves, minced

½ teaspoon cornstarch

½ teaspoon finely grated fresh ginger

4 (6- to 8-ounce) salmon fillets

4 teaspoons vegetable oil

Steamed rice, for serving

1 teaspoon toasted sesame seeds

1. Set your sous vide circulator to 128°F (53.3°C).

2. Mix the soy sauce, sugar, mirin, garlic, cornstarch, and ginger in a small saucepan over medium heat. When it begins to simmer, continue stirring until the sugar has dissolved, about 1 minute. Cool to room temperature and reserve ¼ cup of sauce.

3. Place a maximum of two salmon fillets in each of two food-safe bags and divide the remaining sauce evenly among the bags. Seal the bags using your vacuum sealer or the water displacement method, and place in the refrigerator for at least 1 hour. Place the sealed bags in the water bath and cook for 40 minutes.

4. When the fish is cooked, remove it from the bags and pat dry with paper towels.

5. In a small saucepan over medium-high heat, stir the remaining ¼ cup of sauce for 1 to 2 minutes until it begins to thicken.

continued

6. Heat a large skillet over medium-high heat and add the oil. Gently place the salmon fillets in the skillet skin-side down and cook for 30 seconds. Carefully flip the fillets and cook for another 30 seconds.

7. Remove the fillets, place them on the steamed rice, and spoon the warmed sauce over. Sprinkle with the sesame seeds and serve.

Tip: Brining is important as it prevents the albumin (the white stuff) from leaching out of the fish. It also helps prevent overcooking and results in buttery, moist flesh. Dissolve ¼ cup kosher salt in 1 quart ice water, add the salmon, cover, and refrigerate for 30 minutes.

Tip: An alternative way to firm up the fish is to dry-cure the fillets before bagging them. Sprinkle salt and sugar all over the fillets, let them stand for 10 minutes, then wash the mixture off with cold water.

PESCADO ENCOCADO

Prep time: 15 MINUTES, PLUS 1 HOUR MARINATING TIME | Cook time: 40 MINUTES
Finishing time: 2 MINUTES | 138°F (58.8°C)

I fell in love with this traditional Ecuadorian dish during my honeymoon in South America. Ecuadorian and Peruvian cuisine is heavily centered around fresh fish. Traditionally, most recipes are cooked ceviche, however, I was also fortunate to taste slow-cooked fish dishes with lime, coconut, peppers, and onions. **4 TO 6 SERVINGS**

½ cup freshly squeezed orange juice
¼ cup freshly squeezed lime juice
4 garlic cloves, pressed
1 teaspoon coriander seeds, crushed
1 teaspoon cumin powder
1 teaspoon paprika
½ teaspoon sea salt, divided
4 to 6 fresh halibut or mahi mahi
  fillets, skin removed, cut into
  1 ½- to 2-inch chunks

2 tablespoons vegetable oil
4 Roma tomatoes, peeled and chopped
2 bell peppers, diced
1 yellow onion, diced
14 ounces coconut milk
¼ cup shredded unsweetened coconut
3 tablespoons finely chopped fresh
  cilantro

1. Set your sous vide circulator to 138°F (58.8°C).

2. Mix together the orange juice, lime juice, garlic, coriander, cumin, paprika, and ¼ teaspoon of salt in a small bowl.

3. Add the fillets to the citrus marinade and toss. Cover and store in the refrigerator for 1 hour.

4. Meanwhile, heat the vegetable oil over medium heat in a large saucepan. Add the tomatoes, bell peppers, onion, and the remaining ¼ teaspoon of salt and cook until softened, 4 to 5 minutes.

5. Slowly add the coconut milk to the pan, stirring quickly. Continue stirring and cooking for 10 minutes. Remove from the heat and let cool.

6. Remove the fish from the refrigerator and equally divide the fillets and marinade between two food-safe bags. Divide the coconut milk sauce evenly between the two bags and seal using your vacuum sealer or the water displacement method. Place the sealed bags in the water bath and cook for 40 minutes.

7. Transfer the fish and sauces to a serving dish and sprinkle with the shredded coconut and chopped cilantro. Serve over a bed of rice.

DIVER SCALLOPS *with* CANDIED BACON

Prep time: 5 MINUTES | Cook time: 30 MINUTES | Finishing time: 10 MINUTES | **125°F (51.6°C)**

A different take on a more traditional dish, bacon-wrapped scallops, the scallops in this recipe are cooked simply with salt and pepper, finished in a buttered pan, and topped with candied bacon. The sweetness and smokiness of the bacon can help mask the perceived "fishy" taste sometimes associated with scallops. Serve as an appetizer to share or doubled up as a main. **6 SERVINGS AS AN APPETIZER, 3 SERVINGS AS A MAIN**

12 large fresh diver scallops
1 tablespoon extra-virgin olive oil
¼ teaspoon sea salt
½ teaspoon freshly ground black pepper, divided

4 bacon slices
2 tablespoons brown sugar
2 tablespoons unsalted butter

1. Set your sous vide circulator to 125°F (51.6°C). Preheat the oven to 400°F.

2. In a small bowl, toss the scallops with the olive oil, salt, and ¼ teaspoon of pepper, then place them in a food-safe bag in a single layer and seal using your vacuum sealer or the water displacement method. Place the sealed bag in the water bath and cook for 30 minutes.

3. Meanwhile, lay the bacon on an aluminum foil–lined baking sheet and sprinkle both sides with the brown sugar and the remaining ¼ teaspoon of pepper. (If you prefer not to candy the bacon, omit the sugar and just lay the bacon flat on the foil-lined baking sheet.) Cook the bacon in the oven for 20 minutes. Transfer the bacon to a paper towel–lined plate to remove excess fat. Reserve any bacon fat from the baking sheet for another use.

4. When the scallops are cooked, use tongs to remove them from the bag and place on a paper towel–lined plate. Pat them dry.

5. In a large cast iron pan over medium heat, melt the butter. Add 1 teaspoon of the reserved bacon fat and stir to mix the fats.

6. Add the scallops to the pan and sear until golden brown, 1 minute per side.

7. Cut the bacon into small pieces or shards. Plate the cooked scallops, top with a small stack of candied bacon, and serve.

CITRUS-DILL POACHED HALIBUT

Prep time: 5 MINUTES, PLUS 30 MINUTES MARINATING TIME | Cook time: 30 MINUTES
Finishing time: 10 MINUTES | **137°F (58.3°C)**

Halibut was the first fish I ever made sous vide. It becomes so tender and flaky it can be a challenge to get it out of the bag in one piece, so be careful. The butter, lemon, and dill combine for fresh flavor, and when the skin is crisped up with butter post–sous vide, you've got a perfect and healthy main course. **4 SERVINGS**

4 fresh halibut fillets, skin on
½ teaspoon sea salt
6 tablespoons unsalted butter, divided
Zest of 1 lemon

Juice of 1 lemon
2 teaspoons finely chopped fresh dill
1 lemon, quartered

1. Set your sous vide circulator to 137°F (58.3°C).

2. Lay the fillets on a flat surface and season them all over with the salt, then place two fillets in each of two food-safe bags. To each bag add 2 tablespoons of butter, half the lemon zest and lemon juice, and 1 teaspoon of dill.

3. Seal the bags using your vacuum sealer or the water displacement method. (If using a vacuum sealer, use a gentle setting to manually stop the seal before the fillets get crushed. For a delicate fish such as halibut, I recommend using the water displacement method.) Refrigerate the halibut for 30 minutes to marinate.

4. Place the sealed bags in the water bath and cook for 30 minutes.

5. When the fish is cooked, use a spatula or fish lifter to carefully remove the fillets and place on paper towels. Pat them dry.

6. Add the remaining 2 tablespoons of butter to a large cast iron pan over medium-high heat. Sear the halibut, skin-side down, for 45 seconds, spooning the melted butter over the top. Flip and cook the other side for another 30 to 45 seconds, continuing to spoon butter over the top.

7. Drain the fillets on paper towels before plating, and serve with the lemon quarters.

CANADIAN MAPLE-GLAZED TROUT

Prep time: 15 MINUTES | Cook time: 45 MINUTES | Finishing time: 5 MINUTES | **115°F (46.1°C)**

As a Canadian, I love my maple syrup. It has to be 100 percent pure maple syrup or nothing. Canada also has vast freshwater lakes, packed with all types of trout. Away from the cities, you could find both fresh fish and fresh maple syrup within arm's reach of each other. In this recipe, the sweetness helps mask the powerful fish flavor trout has, keeping it fresh and not overly rich. Serve on a bed of rice or with Garlic and Herb Cheese Cauliflower Purée (page 56) or Garlic-Parmesan Asparagus (page 57). **4 SERVINGS**

6 tablespoons unsalted butter or ¼ cup extra-virgin olive oil, divided

2 tablespoons Dijon mustard

2 tablespoons brown sugar

2 tablespoons pure maple syrup

2 tablespoons low-sodium soy sauce

2 tablespoons white wine vinegar

4 fresh rainbow or steelhead trout fillets, skin on

1. Set your sous vide circulator to 115°F (46.1°C).

2. In a small saucepan over medium heat, combine 4 tablespoons of butter, the mustard, brown sugar, maple syrup, soy sauce, and vinegar, and stir until the butter has melted.

3. Remove from the heat and let cool for 5 minutes, stirring occasionally.

4. Place two trout fillets in each of two food-safe bags, evenly divide the butter-maple sauce between the two bags, and seal using your vacuum sealer or the water displacement method. Place the sealed bags in the water bath and cook for 45 minutes.

5. When the fish is cooked, use a spatula or fish lifter to carefully remove the fillets and place on paper towels. Pat them dry.

6. Melt the remaining 2 tablespoons of butter in a cast iron pan over medium-high heat. Gently sear the fillets, skin-side down, for 1 to 2 minutes.

LOBSTER ROLLS

Prep time: 5 MINUTES | Cook time: 50 MINUTES | Finishing time: 5 MINUTES | 137°F (58.3°C)

If you want to factor lobster into a lunchtime dish, the lobster roll is the perfect way to do that. You still get your perfectly cooked lobster tails, but combined with the fresh celery and parsley, you'll impress your lunchtime guests. **4 SERVINGS**

4 medium to large lobster tails,
   in the shell
2 garlic cloves, pressed
Zest of ½ lemon
½ cup freshly squeezed lemon
   juice, divided
4 tablespoons unsalted butter, divided

½ cup mayonnaise
2 celery stalks, chopped
2 tablespoons chopped fresh parsley
⅛ teaspoon sea salt
⅛ teaspoon freshly ground black pepper
4 rolls, split, lightly buttered, and toasted

1. Set your sous vide circulator to 137°F (58.3°C).

2. Bring a large pot of water to a boil. Add the lobster tails to the water and parboil for 90 seconds. Remove the tails and set aside to cool.

3. Mix together the garlic, lemon zest, and ¼ cup of lemon juice in a small bowl.

4. When cool to the touch, squeeze the tails inward from the edges to crack the tails down the center of the shell. Carefully remove the meat from the tails.

5. To each of two food-safe bags, add equal amounts of lobster meat, 2 tablespoons of butter and half of the lemon-garlic mix, and seal using your vacuum sealer or the water displacement method. Place the sealed bags in the water bath and cook the tails for 50 minutes.

6. When the lobster is cooked, remove the bags from the water bath, remove the tail meat from the cooking bags, and discard the liquid.

7. Using a sharp knife, cut the lobster meat into 1-inch pieces and gently mix with the remaining ¼ cup lemon juice, mayonnaise, celery, parsley, salt, and pepper in a bowl.

8. Spoon the lobster mix into the rolls, almost overfilling them, and serve.

CAJUN SHRIMP SKEWERS

Prep time: 15 MINUTES | Cook time: 30 MINUTES | Finishing time: 5 MINUTES | 137°F (58.3°C)

Easy to make and packed full of flavor, these shrimp skewers are a perfect appetizer, or you can double them up for a main. Cajun spices and shrimp have been married for a long time, often in creoles or gumbos. This recipe is a quick way get some Cajun flavors on the table without having to wait hours to get it! Make sure you squeeze some fresh lemon juice on to help cut some of the initial heat. **4 SERVINGS**

6 tablespoons unsalted butter

2 garlic cloves, pressed

Zest of 1 lemon

1½ tablespoons freshly squeezed lemon juice

1 teaspoon kosher salt

1 teaspoon freshly ground black pepper

1 tablespoon Cajun spice powder

1½ pounds jumbo shrimp, peeled and deveined, tails on

1 tablespoon minced fresh parsley, plus additional for garnish (optional)

Lemon wedges

1. Set your sous vide circulator to 137°F (58.3°C). Fill a large bowl with water and soak 8 skewers while you cook the shrimp.

2. Heat the butter, garlic, lemon zest and juice, salt, pepper, and Cajun spice powder in a medium saucepan over low heat until the butter is melted, about 5 minutes. Stir and mix well then let cool.

3. Add the shrimp to a food-safe bag and then add the melted butter mixture to the bag, massaging the shrimp with the sauce to ensure full coverage.

4. Flatten the shrimp into a single layer before sealing with your vacuum sealer or the water displacement method. Massage again to ensure the shrimp are in a single layer. Place the sealed bag in the water bath and cook for 30 minutes.

5. When the shrimp are cooked, transfer them to a paper towel–lined plate and discard the cooking liquid.

6. Add 4 shrimp to each skewer, sprinkle with the parsley, squeeze fresh lemon across them, and serve.

FISH TACOS

Prep time: 5 MINUTES | Cook time: 25 MINUTES | Finishing time: 5 MINUTES | 135°F (57.2°C)

Fish tacos have become increasingly popular over the past few years. Traditionally served grilled or fried, they combine so well with the fresh citrus, chile, and cilantro flavors, your guests will be coming back for second and third helpings. The fish is very tender and flaky, but when torched before serving, it picks up the grilled flavor that makes the perfect fish taco. **4 TO 6 SERVINGS**

4 fillets cod or mahi mahi, skinned

1 teaspoon chopped fresh cilantro

¼ teaspoon red pepper flakes

½ teaspoon kosher salt, divided

½ teaspoon freshly ground black pepper, divided

1 tablespoon cider vinegar

½ red onion, chopped

4 to 6 corn or flour tortillas

Shredded iceberg lettuce

1 large tomato, sliced

⅓ cup sour cream

Guacamole, purchased or homemade

1 lime, quartered

1. Set your sous vide circulator to 135°F (57.2°C).

2. Lay the fish fillets on a flat surface and season with the cilantro, red pepper flakes, ¼ teaspoon of salt, and ½ teaspoon of pepper.

3. Add the fillets to a food-safe bag and seal with the gentle mode of your vacuum sealer or using the water displacement method. Add the sealed bag to the water bath and cook for 25 minutes.

4. Meanwhile, combine the cider vinegar, onion, the remaining ½ teaspoon of salt, and the remaining ½ teaspoon of pepper in a small bowl and set aside.

5. When the fish is cooked, use a spatula or fish lifter to carefully remove the fillets and place on paper towels. Pat them dry.

6. Using a blowtorch, lightly sear the fish before cutting into smaller chunks.

7. To serve, into a warmed tortilla, spoon a portion of cooked fish, some lettuce, tomato, sour cream, onion mixture, guacamole, and a good squeeze from a lime wedge.

GRILLED DEEP-SEA SWORDFISH

Prep time: 5 MINUTES | Cook time: 35 MINUTES | Finishing time: 5 MINUTES | 125°F (51.6°C)

I'll never forget the first time I ate swordfish: 1988 in Ixtapa, Mexico. It was my first deep-sea fishing excursion. After a slow morning, my rod suddenly dipped. I climbed into the chair and, feeling the tug, knew I had something very large on the line. Aware that my seven-year-old frame was no match for the fish, my dad strapped in and began the fight. Two hours later an eight-foot swordfish lay across the back of the boat, so long it hung over the edges. We donated the fish to the resort restaurant, and they served it for dinner that evening. **4 TO 6 SERVINGS**

4 (6- to 8-ounce) fresh swordfish steaks
½ teaspoon sea salt
¼ teaspoon freshly ground black pepper
6 tablespoons extra-virgin
  olive oil, divided
1 tablespoon champagne vinegar

1 tablespoon freshly squeezed
  lemon juice
Zest of 1 lemon
2 garlic cloves, pressed
1 teaspoon chopped fresh basil
1 teaspoon chopped fresh rosemary
1 teaspoon chopped fresh thyme

1. Set your sous vide circulator to 125°F (51.6°C).

2. Lay the swordfish steaks on a clean flat surface and season with the salt and pepper.

3. Whisk together 4 tablespoons of olive oil, vinegar, lemon juice and zest, garlic, basil, rosemary, and thyme in a small bowl.

4. Place a maximum of two steaks into each of two food-safe bags, pour the citrus marinade evenly over the swordfish, and seal using your vacuum sealer or the water displacement method. Place the sealed bags in the water bath and cook for 35 minutes.

5. When the swordfish is cooked, remove the steaks from the bag (reserving the cooking liquid) and place on a paper towel–lined plate or baking sheet. Pat them dry.

6. Heat a cast iron pan over high heat and add the remaining 2 tablespoons of olive oil. When the oil is very hot, gently place the steaks in the pan and cook for 1 minute per side, rotating 90 degrees every 30 seconds.

7. Remove the steaks and plate. Drizzle the reserved cooking liquids over the fish and serve with rice, kale, or Garlic-Parmesan Asparagus (page 57).

SEARED WASABI AHI TUNA

Prep time: 5 MINUTES, PLUS 15 MINUTES RESTING TIME | Cook time: 30 MINUTES
Finishing time: 5 MINUTES | **105°F (40.5°C)**

I avoided eating tuna for way too long. I associated it with canned tuna and I just couldn't bring myself to eat it. Fast-forward to my early thirties, when I ate fresh tuna sashimi for the first time and I left the restaurant thinking "what took me so long?" I've made up for lost tuna time over the years, and I enjoy it every chance I get. Cooking tuna sous vide adds firmness to the tuna that creates a middle ground between sashimi-raw and fully cooked tuna. **4 SERVINGS**

2 (8-ounce) ahi tuna steaks
½ teaspoon sea salt
½ teaspoon freshly ground black pepper

2 tablespoons extra-virgin olive oil
½ teaspoon wasabi powder
2 teaspoons canola oil

1. Set your sous vide circulator to 105°F (40.5°C).

2. Lay the tuna steaks on a clean flat surface and season with the salt and pepper.

3. Whisk together the olive oil and wasabi powder in a small bowl.

4. Place the steaks in a single layer in a food-safe bag, pour the wasabi oil over, and seal using your vacuum sealer or the water displacement method. Let rest in the refrigerator for 15 minutes. Add the sealed bag to the water bath and cook for 30 minutes.

5. Heat a cast iron pan over high heat and add the canola oil.

6. Remove the steaks from the bag, transfer to a paper towel–lined plate or baking sheet, and pat them dry. Discard the cooking liquid.

7. Gently place the steaks in the pan and sear for 30 seconds per side.

8. Slice the steaks into 1-inch thick slices, and serve.

NIÇOISE SALAD

Prep time: 10 MINUTES | Cook time: 2 HOURS, 45 MINUTES | Finishing time: 10 MINUTES
TUNA STEAK: 105°F (40.5°C), EGGS: 167°F (75°C), POTATOES: 190°F (87.8°C)

I was looking for a recipe that combined multiple sous vide elements—and was surprised to find it was a salad, of all things. Adapting the potato and tuna recipes from this book made them great candidates to perfect this dish and blow away your friends with your sous vide salad. **2 SERVINGS**

FOR THE TUNA
1 (8-ounce) ahi tuna steak
¼ teaspoon sea salt
¼ teaspoon freshly ground black pepper
1 tablespoon extra-virgin olive oil
1 teaspoon canola oil

FOR THE EGGS
3 eggs

FOR THE POTATOES
1 pound russet or Yukon gold potatoes
3 tablespoons extra-virgin olive oil
2 teaspoons sea salt
2 or 3 fresh thyme sprigs

FOR THE SALAD
1 cup grape tomatoes
1/4 cup Niçoise olives
1 tablespoon extra-virgin olive oil
1 teaspoon Dijon mustard
3 tablespoons cider vinegar
1/4 teaspoon kosher salt
6 ounces fresh arugula leaves

TO MAKE THE TUNA

1. Set your sous vide circulator to 105°F (40.5°C).

2. Lay the tuna steak on a clean flat surface and season with the sea salt and pepper.

3. Place the steak in a single layer in a food-safe bag, add the olive oil, and seal using your vacuum sealer or the water displacement method. Place the bag in the water bath and cook for 30 minutes.

4. Heat a cast iron pan over high heat and add the canola oil.

5. Remove the steak from the bag, transfer to a paper towel–lined plate, and pat dry. Discard the cooking liquid.

6. Gently place the steak in the pan and sear for 30 seconds per side.

7. Slice the steak into 1-inch-thick slices. Set aside and cover with plastic wrap or aluminum foil.

TO MAKE THE EGGS

1. Set your sous vide circulator to 167°F (75C°).

2. Gently lower the eggs directly into the water bath and cook for 1 hour.

3. When cooked, remove the eggs and let cool before peeling.

4. Slice the eggs in half. Transfer to a bowl, cover, and place in the refrigerator.

TO MAKE THE POTATOES

1. Set your sous vide circulator to 190°F (87.8°C).

2. Add the potato, olive oil, sea salt, and thyme to a food-safe bag and seal using your vacuum sealer or the water displacement method.

3. Place the sealed bag in the water bath and cook for 1 hour, 15 minutes.

4. Remove the bag from the water bath, but do not open it. Place it on the countertop if making the salad immediately, or store in the refrigerator for 1 to 2 days. (If using refrigerated potatoes, either place the sealed bag in the sous vide bath while the tuna cooks, or set your sous vide circulator to 150°F [65.5°C] and place the bag in the water bath for 30 minutes to warm the potatoes.)

TO MAKE THE SALAD

1. Halve the grape tomatoes, and pit the olives and cut into quarters.

2. Whisk together the olive oil, mustard, cider vinegar, and salt in a large bowl.

3. Add the arugula, tomatoes, and olives to the dressing and toss to mix.

4. Divide the salad base between two plates. Divide the egg halves equally between the plates. Remove the potatoes from the bag, discard the cooking liquid, and cut the potatoes in half. Top the salad with the potatoes and the tuna.

ICELANDIC LEMON-DIJON ARCTIC CHAR

Prep time: 5 MINUTES | Cook time: 20 MINUTES | Finishing time: 5 MINUTES | 115°F (46.1°C)

For my thirty-fifth birthday, my wife surprised me with a trip to Reykjavik, Iceland. If you haven't been, I can't recommend it enough. Iceland is the world's largest producer of arctic char, by way of the land-based fish farms, and they also know how to cook it very well. The cuisine in Reykjavik is one of the many reasons I plan on returning. Serve this arctic char with Asiago Broccoli and Peppers (page 58) or over a bed of rice with freshly chopped tomatoes. **4 SERVINGS**

4 (6-ounce) arctic char fillets, skin-on
½ teaspoon sea salt
½ teaspoon freshly ground black pepper
Zest of 1 lemon

Juice of 1 lemon, divided
1 tablespoon extra-virgin olive oil
2 teaspoons Dijon mustard
2 teaspoons canola oil

1.   Set your sous vide circulator to 115°F (46.1°C).

2.   Lay the fish fillets on a clean flat surface and season with the salt and pepper.

3.   Combine the zest, juice of half a lemon, oil, and mustard into a small bowl and whisk until thoroughly mixed.

4.   Place two fillets in a single layer in each of two food-safe bags, pour in the lemon-mustard mixture, and seal using your vacuum sealer or the water displacement method. Place the sealed bags in the water bath and cook for 20 minutes.

5.   Heat a cast iron pan over high heat and add the canola oil.

6.   Remove the fillets from the bag, transfer to a paper towel–lined plate or baking sheet, and pat them dry. Discard the cooking liquids.

7.   Gently place the fillets in the pan and sear for 30 seconds per side. Turn off the heat and add the other half of the lemon juice.

SWEET UMAMI SALMON BELLY

Prep time: 5 MINUTES | Cook time: 30 MINUTES | Finishing time: 0 MINUTES | **130°F (54.4°C)**

Salmon belly, also known as sake toro, is the fatty part of the salmon. It is rich, fat, and full of flavor. It is often served in higher-end Japanese restaurants and can be expensive compared to traditional salmon sushi or sashimi. In my opinion it is worth the extra cost, and anytime I see salmon or tuna belly on a menu, I can't resist ordering it. Now I can enjoy it in the comfort of my own home with this simple sous vide recipe! **4 SERVINGS**

2 (8-ounce) fillets salmon belly, skinned
1 teaspoon sea salt
¼ teaspoon freshly ground black pepper
½ cup soy sauce, divided

¼ cup light brown sugar
¼ cup extra-virgin olive oil
2 garlic cloves, pressed
1 tablespoon wasabi

1. Set your sous vide circulator to 130°F (54.4°C).

2. Lay the salmon belly on a flat surface and season with the salt and pepper.

3. Combine the soy sauce, brown sugar, olive oil, and garlic in a small bowl and whisk until thoroughly mixed.

4. Place the fillets in a single layer in a food-safe bag, pour in the garlic-soy mixture, and seal using your vacuum sealer or the water displacement method. Place the sealed bag in the water bath and cook for 30 minutes.

5. Remove the fillets from the bag, transfer to a paper towel–lined plate or baking sheet, and pat them dry. Discard the cooking liquid.

6. Serve immediately over rice or sliced with soy sauce and wasabi as dipping sauces.

LEMON-PARSLEY CUTTLEFISH SPAGHETTI

Prep time: 5 MINUTES | Cook time: 2 HOURS | Finishing time: 5 MINUTES | **135°F (57.2°C)**

Cuttlefish has flavors similar to calamari and octopus, and its texture and consistency is tough, making it an excellent candidate for sous vide cooking. This is a simple dish to prepare and will reward your sense of curiosity and adventure with a fresh and light pasta dish. **4 SERVINGS**

4 (6-ounce) cuttlefish bodies, cleaned

1 teaspoon sea salt

½ teaspoon freshly ground black pepper

3 tablespoons extra-virgin olive oil, divided

10 ounces dried spaghetti

1 (16-ounce) can Roma tomatoes

2 garlic cloves, pressed

1 teaspoon red pepper flakes

Zest of 1 lemon

1 tablespoon freshly squeezed lemon juice

3 tablespoons chopped fresh parsley, divided

1.   Set your sous vide circulator to 135°F (57.2 °C).

2.   Lay the cuttlefish on a clean flat surface and season with the salt and pepper.

3.   Place the cuttlefish and 2 tablespoons of olive oil in a food-safe bag in a single layer. Seal using your vacuum sealer or the water displacement method. Place the sealed bag in the water bath and cook for 2 hours.

4.   After 1 hour, 45 minutes, bring a large pot of water to the boil and cook the spaghetti until al dente. Drain.

5.   Add the remaining 1 tablespoon of olive oil, tomatoes, garlic, red pepper flakes, lemon zest, lemon juice, and 2 tablespoons of parsley to a medium skillet over medium heat and sauté for 3 minutes. Reduce the heat to a simmer.

6. When the cuttlefish is cooked, remove them from the bag, place on a paper towel–lined plate, and pat them dry. Discard the cooking liquid.

7. Lay the fish on a clean flat surface and cut into 1- to 2-inch pieces.

8. Combine the cooked pasta, tomato sauce, and cuttlefish in the skillet and toss until thoroughly mixed.

9. Sprinkle the remaining 1 tablespoon of parsley over the pasta and serve immediately.

JUMBO PRAWNS *with* HOMEMADE COCKTAIL SAUCE

Prep time: 5 MINUTES, PLUS 1 HOUR COOLING TIME | Cook time: 30 MINUTES
Finishing time: 5 MINUTES | **137°F (58.3°C)**

This cocktail party classic is perfect for sous vide cooking because it is easy to make, can be prepared ahead of the evening, and can scale up to meet your guest count. Your guests will be sure to love this combination, especially if you let them know that you made the cocktail sauce yourself. **4 TO 6 SERVINGS**

16 to 20 jumbo prawns, deveined and
  peeled, tails on
1 cup ketchup
2 tablespoons horseradish
1 teaspoon Worcestershire sauce

1 teaspoon freshly squeezed lemon juice
1 teaspoon Tabasco sauce
¼ teaspoon kosher salt
⅛ teaspoon freshly ground black pepper

1. Set your sous vide circulator to 137°F (58.3°C).

2. Place the prawns in a single layer in a large food-safe bag, or divide between two smaller bags. Seal using your vacuum sealer or the water displacement method. Place the sealed bag in the water bath and cook for 30 minutes.

3. When the prawns are cooked, place the bag in an ice bath for 10 minutes then transfer to the refrigerator for at least 1 hour and no more than 6 hours.

4. Combine the ketchup, horseradish, Worcestershire sauce, lemon juice, Tabasco, salt, and pepper in a small bowl and whisk until thoroughly mixed. Cover with plastic wrap and store in the refrigerator until ready to serve.

DOVER SOLE *with* LEMON HERBED BUTTER

Prep time: 5 MINUTES | Cook time: 30 MINUTES | Finishing time: 2 MINUTES | 134°F (56.6°C)

A very delicate fish and one that is so easily overcooked, Dover sole is amazingly tender when cooked correctly. Traditionally served with lemon herbed butter, here I use the same water bath to prepare this herbal melted goodness so everything is ready at the same time. **6 SERVINGS**

6 (6-ounce) sole fillets
1 teaspoon kosher salt
1 teaspoon freshly ground black pepper
Zest of ½ lemon
8 tablespoons unsalted butter

1 tablespoon minced fresh
  parsley, divided
½ tablespoon minced fresh chives
½ tablespoon minced fresh basil
½ tablespoon minced fresh thyme

1. Set your sous vide circulator to 134°F (56.6°C).

2. Lay the fillets on a clean flat surface and season with the salt and pepper.

3. Place the fillets and lemon zest in a single layer in a food-safe bag and seal using your vacuum sealer or the water displacement method.

4. In a separate food-safe bag, combine the butter, ½ tablespoon of parsley, chives, basil, and thyme, and seal using your vacuum sealer or the water displacement method. Place both sealed bags in the water bath and cook for 30 minutes.

5. When the fish is cooked, remove the fillets, place them on a paper towel–lined plate or baking sheet, and pat them dry. Discard any cooking liquid.

6. Remove the herbed butter bag, open it, and spoon the butter over the fillets. Sprinkle the remaining ½ tablespoon of parsley over the fish and serve.

chapter five
Poultry and Pork

Poultry and pork dishes are an area where sous vide cooking, with its ability to retain moisture during the cooking process, brings out some of the best results. We've all had dry and overcooked chicken breasts, or dry and tough ribs. Poultry and pork dry out during traditional cooking due to the evaporation of moisture, whether natural or absorbed from a marinade. When poultry and pork are sealed for sous vide, not only are their flavors and aromatics sealed in, but cooking them sous vide also ensures that their juices and moisture don't escape. Here are some recipes so you can enjoy the juiciest chicken breasts, the moistest pork tenderloins, and a few specialty dishes that work very well with sous vide cooking.

LIME AND GARLIC PORK TENDERLOIN (page 109)

DUCK CONFIT

Prep time: 15 MINUTES, PLUS 12 HOURS COOLING TIME | Cook time: 12 HOURS
Finishing time: 5 TO 7 MINUTES | **170°F (76°C)**

This classic French dish is prepared using a traditional preservation process combining salt curing and cooking in fat. If stored properly (in a vacuum-sealed bag), duck confit lasts for up to two weeks, so can be prepared in advance then finished when ready to serve. It's delicious alongside a salad of mixed greens with a mustard vinaigrette, some crusty French bread, and a bottle of good French wine. **6 SERVINGS**

6 duck legs	1 fresh rosemary sprig, stemmed
10 teaspoons ground coriander	1½ cups duck fat
10 teaspoons sea salt	3 bay leaves
6 garlic cloves, smashed	1 teaspoon peppercorns
9 fresh thyme sprigs	Zest of 1 orange (optional)

1. Twelve to 24 hours before cooking, place the duck in a dish. Mix together the coriander and salt and rub the mixture into the duck legs. Scatter the garlic, thyme, and rosemary over the duck. Cover and refrigerate for 12 to 24 hours.

2. Set your sous vide circulator to 170°F (76°C).

3. Remove the duck from the refrigerator and pat it dry. Do not wash the duck at this stage; it is important to preserve the salt curing before the cooking process.

4. Add the duck legs, duck fat, bay leaves, peppercorns, and orange zest (if using) to a food-safe bag and seal using your vacuum sealer or the water displacement method. Place the sealed bag in the water bath and cook for 12 hours. Cover the water bath to avoid excess evaporation, and check periodically to ensure the water level remains sufficient to cover the bag at all times.

5. When the duck is cooked, if you are not serving it immediately, leave it in the sealed bag and refrigerate for up to 10 days.

6. When ready to serve, remove the duck from the bag, wipe excess fat off the duck, and heat a cast iron pan over high heat for 1 minute. Add the duck legs and sear until golden brown and crispy all over, 5 to 7 minutes.

Tip: Goose fat can be substituted for duck fat. Whichever you use, after cooking, the fat can be strained, cooled, and reused for another recipe such as Crispy-Skin Smashed Potatoes (page 49).

DUCK BREAST

Prep time: 15 MINUTES | Cook time: 1 HOUR, 30 MINUTES | Finishing time: 5 MINUTES | 144°F (62°C)

The key to a delicious duck breast is ensuring the fat is rendered properly and the skin is nice and crispy, which is why in this recipe it gets seared twice. Though this recipe is simple, duck pairs beautifully with many different flavors including orange, maple, lingonberry, plum, and rosemary. You can try each of them out; just add to the sous vide bag after searing the duck breast. **2 SERVINGS**

2 boneless duck breasts, skin on
Sea salt
Freshly ground black pepper

3 fresh thyme sprigs
2 juniper berries, crushed (optional)
1 bay leaf

1. Set your sous vide circulator to 144°F (62°C).

2. Place the duck on a cutting board, skin-side up, and score the skin in a crosshatch pattern with a sharp knife, being careful not to cut through the fat into the meat.

3. Heat a cast iron pan over medium-high heat for 2 minutes. Place the duck skin-side down in the pan and sear until the skin is a rich golden brown, about 5 minutes. Use a spatula to press the duck breasts down gently to ensure the skin sears evenly. Remove the duck from the pan and reserve any rendered fat.

4. Liberally season the duck breasts with salt and pepper. Add the duck, thyme, juniper berries (if using), bay leaf, and reserved rendered fat to a food-safe bag and seal using your vacuum sealer or the water displacement method. Place the sealed bag in the water bath and cook for 1 hour, 30 minutes.

5. Heat a cast iron pan over high heat for 3 minutes. Remove the duck from the bag, place it skin-side down in the pan, and sear until the skin is rich golden brown, about 5 minutes, pressing down to ensure an even sear.

NO-GRILL SUMMER SAUSAGES

Prep time: 5 MINUTES | Cook time: 45 MINUTES | Finishing time: 5 MINUTES | 160°F (71°C)

This recipe is great if you don't have access to a grill, or if you are looking for a little bit of summer on a winter's night and don't fancy standing outside grilling in the snow. Alternatively, if you are grilling up a storm and space on the grill is at a premium, you can always throw the sausages in the sous vide to free up some space. Try the Sous Vide Buttered Corn on the Cob (page 55) for a perfect side dish. **4 TO 6 SAUSAGES**

4 to 6 good quality raw pork sausages
4 to 6 tablespoons apple cider (with or
   without alcohol) (optional)
¾ teaspoon vegetable oil

1.   Set your sous vide circulator to 160°F (71°C).

2.   Add the sausages and 1 tablespoon of cider per sausage (if using) to a food-safe bag and seal using the water displacement method. Place the sealed bag in the water bath and cook for 45 minutes.

3.   Heat the oil in a cast iron pan over medium-high heat for 2 minutes. Add the sausages and cook for 3 to 4 minutes, turning periodically to ensure they brown evenly.

Tip: Because the sausages are very soft when raw, using the water displacement method rather than a vacuum sealer is recommended to help preserve the original shape of the sausage and avoid warping.

ZESTY LEMON-HERB PORK CHOPS

Prep time: 10 MINUTES | Cook time: 1 HOUR | Finishing time: 5 MINUTES | **140°F (60°C)**

This is a flavorful, lean, and versatile dish. For a casual summer dinner, pair it with a fresh salad of garden greens, heirloom tomatoes, hothouse cucumbers, cubed feta, and a light vinaigrette-style dressing. For something more comforting on a chilly night, serve with the Garlic and Herb Cheese Cauliflower Purée (page 56) and Crispy-Skin Smashed Potatoes (page 49). **4 SERVINGS**

4 boneless pork loin chops
Sea salt
Freshly ground black pepper
Zest of 1 lemon
¼ cup freshly squeezed lemon juice

2 garlic cloves, smashed
2 bay leaves
1 fresh oregano sprig
4 tablespoons unsalted butter, divided

1. Set your sous vide circulator to 140°F (60°C).

2. Season the pork chops generously with salt and pepper.

3. Add the pork chops, lemon zest and juice, garlic, bay leaves, oregano, and 2 tablespoons of butter to a food-safe bag and seal using your vacuum sealer or the water displacement method. Place the sealed bag in the water bath and cook for 1 hour.

4. Remove the pork chops from the bag and pat them dry with a paper towel. Reserve the herbs from the bag.

5. Heat the remaining 2 tablespoons of butter and the reserved herbs in a cast iron pan over medium-high heat for 2 minutes. Add the pork chops and sear on each side until golden brown, 1 to 1½ minutes. Do not crowd the pan; cook in batches if necessary.

CHICKEN *and* CHORIZO TACOS

Prep time: 45 MINUTES, PLUS 1 HOUR MARINATING TIME | Cook time: 1 HOUR, 30 MINUTES
Finishing time: 10 MINUTES | **140°F (60°C)**

Whether you love spicy food or prefer milder flavors, you can dial the spice level on this dish up or down. For milder tastes, omit the jalapeño altogether, and for more spice add more jalapeños to taste. **8 TACOS**

½ cup water
1 poblano pepper, stemmed and seeded
½ jalapeño pepper, stemmed and seeded (optional)
4 scallions, chopped
Leaves of 1 bunch fresh cilantro
½ cup chopped fresh parsley
3 garlic cloves
2 tablespoons freshly squeezed lime juice
1 teaspoon sea salt
¾ teaspoon ground coriander

¾ teaspoon ground cumin
4 skinless boneless chicken breasts, cut into strips
2 pork chorizo sausages
1 tablespoon vegetable oil
½ yellow onion, sliced thinly
8 hard or soft corn taco shells
Queso fresco or shredded Cheddar cheese (optional)
Diced fresh tomato (optional)
Shredded iceberg lettuce (optional)

1. Place the water, poblano, jalapeño (if using), scallions, cilantro, parsley, garlic, lime juice, salt, coriander, and cumin in a blender and blend until smooth.

2. Place the chicken strips and cilantro mixture in a food-safe bag and seal using your vacuum sealer or the water displacement method. Put the bag in the refrigerator to marinate for at least 1 hour.

3. Set your sous vide circulator to 140°F (60°C).

4. Remove the bag from the refrigerator, place it in the water bath, and cook for 1 hour, 30 minutes. 20 minutes before the end of the cooking time, remove the casings from the chorizo sausages and break up the meat with your hands.

5. Add the vegetable oil to a nonstick skillet and heat over medium-high heat for 2 minutes. Sauté the onion until caramelized, about 3 minutes. Add the chorizo to the onion and stir occasionally until the chorizo is cooked through, 5 to 7 minutes.

6. Remove the chicken from the bag and discard the cooking liquid. Add the chicken to the chorizo mixture and stir until well combined and heated through.

7. Fill each of the taco shells with some of the chicken-chorizo filling and garnish with cheese, tomato, and lettuce (if using).

PORK CARNITAS *with* SALSA FRESCA

Prep time: 10 MINUTES, PLUS 1 HOUR MARINATING TIME | Cook time: 48 HOURS
Finishing time: 30 MINUTES | 158°F (70°C)

Carnitas literally means "little meats" in Spanish and is one of the more tradi-
tional Mexican dishes. The meat is crispy on the outside and juicy and tender on
the inside and is slowly cooked in lard over the course of a few hours. I have light-
ened up the dish a little by removing the lard, although the longer cook time still
achieves the soft melt-in-the-mouth texture you would expect. I suggest serving
this with salsa fresca, but carnitas would be equally delicious with refried beans,
guacamole, or corn salsa. **8 TACOS**

2 tablespoons red pepper flakes
3 teaspoons kosher salt, plus more
   for seasoning
2 teaspoons chipotle chili powder
2 teaspoons dried oregano
½ teaspoon ground cinnamon
2¼ pounds boneless pork shoulder
   or pork butt

4 small ripe tomatoes, diced
¼ red onion, finely diced
¼ cup fresh cilantro leaves,
   finely chopped
Freshly squeezed lime juice
3 tablespoons vegetable oil
8 corn tortillas

1. Mix together the red pepper flakes, kosher salt, chili powder, oregano, and cinnamon in a small bowl.

2. Coat the pork all over with the rub. Cover with aluminum foil and leave at room temperature to marinate for 1 hour.

3. Set your sous vide circulator to 158°F (70°C).

4. Put the pork in a food-safe bag and seal using your vacuum sealer or the water displacement method. Place the sealed bag in the water bath and cook for 48 hours. Cover the water bath to avoid excessive evaporation, and top up the water periodically to ensure the bag is covered at all times.

5. Fifteen minutes before the end of the cook time, combine the tomatoes, onion, and cilantro in a bowl. Add the lime juice and salt until the mixture has your preferred balance of tartness and saltiness.

6. Remove the bag from the water bath and transfer the pork to a cutting board. Discard any cooking liquid. Using two forks, pull apart the meat until it is finely shredded.

7. Add the vegetable oil to a cast iron pan and heat over medium heat for 3 minutes. Add the shredded pork to the pan and cook, stirring only occasionally to allow the meat to brown and develop crispy and crusty parts.

8. Fill each tortilla with carnitas and top with the salsa fresca.

SRIRACHA HONEY-GARLIC CHICKEN

Prep time: 10 MINUTES | Cook time: 1 HOUR, 30 MINUTES | Finishing time: 5 MINUTES | **140°F (60°C)**

This recipe is all about sweet heat—the sweetness of the raw sugar or honey helps cut some of the spice from the sriracha, making this a good dish for those who want just a little bit of heat. One of our favorite summertime meals, it's also great the next day for lunch, tossed in a fresh salad, or as a snack all on its own. **4 SERVINGS**

8 tablespoons butter

8 garlic cloves, chopped

6 tablespoons sriracha

4 tablespoons raw sugar or honey

Juice of 1 lime

Sea salt

Freshly ground black pepper

4 boneless, skinless chicken breasts

1. Set your sous vide circulator to 140°F (60°C)

2. In a small saucepan over medium heat, simmer the butter, garlic, sriracha, sugar, lime, and a pinch of salt and pepper for 5 minutes. Remove from the heat and set aside to cool.

3. Liberally season the chicken breasts on both sides with salt and pepper and put one chicken breast in each of four food-safe bags.

4. When the marinade has cooled, divide the sauce evenly among all four bags. Seal each bag using your vacuum sealer or the water displacement method. Place the bags in the water bath and cook for 1 hour, 30 minutes.

5. Remove the chicken breasts from the bags and pat them dry.

6. Heat half the liquid from each bag in a large saucepan over medium-high heat.

7. When the sauce is simmering, carefully place the chicken breasts in the pan and cook for 4 minutes.

8. Remove the breasts from the heat, slice, and serve over rice or salad.

Tip: When sealing chicken, or other moist and juicy foods, fold the edge of the top of the bag over to make a cuff. It will help reduce the amount of liquid that can cause you problems with your sealer. Also, if there is a lot of liquid in the bag, pull your sealer close to the counter edge and allow the bag to hang over the edge—this will help prevent the sealer from sucking in too much liquid and not enough air.

SOUS VIDE PULLED PORK *with* CRISPY ONION STRINGS

Prep time: 20 MINUTES | Cook time: 48 HOURS | Finishing time: 20 MINUTES | **158°F (70°C)**

A favorite in my kitchen, this long-cooking recipe is worth the wait. While it takes a little bit of effort, it's simple enough to perfect, and you can be sure that the results will be the same each time you make the dish. I pair it with crispy onion strings to add some texture to the succulent pork. I've used this pulled pork in sandwiches, sliders, and as breakfast, using the pork as a bed for a sous vide poached egg (see page 39). **16 SERVINGS**

FOR THE PORK

4 tablespoons yellow mustard

2 tablespoons soy sauce

2 tablespoons honey

1 tablespoon ketchup

2¼ pounds pork shoulder or pork butt

FOR THE ONIONS

1 large yellow onion, cut into thin rings

2 cups whole milk

1½ cups all-purpose flour

2 teaspoons granulated onion powder

1 teaspoon paprika

Salt

Freshly ground black pepper

4 cups vegetable oil, for frying

TO MAKE THE PORK

1. Set your sous vide machine to 158°F (70°C)

2. Mix together the mustard, soy sauce, honey, and ketchup in a medium bowl to make a paste. Evenly and liberally, coat the entire surface of the pork with the paste. (For extra flavor, you can set aside some of the paste and inject the pork using a food-safe syringe in addition to coating the surface.)

3. Put the meat in a large food-safe bag with any excess paste, making sure the meat is flattened out, and seal using your vacuum sealer or the water displacement method. Place the bag in the water bath and cook for 48 hours.

continued

TO MAKE THE ONIONS

1. Thirty minutes before the end of the cooking time, separate the onion rings with your fingers and place them in a medium bowl. Pour the milk over them and let them sit for about 1 hour.

2. In a separate medium bowl, mix together the flour, onion powder, paprika, and a pinch of salt and pepper.

3. Heat the oil to 375°F.

4. Take a handful of the onions at a time, shake off the excess milk, and dredge them in the flour, making sure you shake off a lot of the flour. Add the onions to the hot oil.

5. Fry them for 2 minutes, stirring after 1 minute to make sure every strand gets crispy.

6. Use a slotted spoon to remove the strands from the oil and place them on a paper towel–lined plate to drain. Repeat with the remaining onions.

TO FINISH THE DISH

1. When the pork is cooked, remove it from the bag and place on a cutting board. Reserve the cooking liquid. Using two forks or a shredder, pull apart the pork.

2. Add the reserved liquid to a small saucepan over medium heat and reduce until sticky, about 5 minutes.

3. Pour the sauce over the pork and serve with the crispy onions.

Tip: To inject flavor right into the pork, use a food-safe cooking syringe, also known as a meat injector. They have a larger gauge needle, are designed to penetrate tougher cuts of meat, allow for thicker liquids to be used, and are dishwasher-safe.

LIME *and* GARLIC PORK TENDERLOIN

Prep time: 15 MINUTES | Cook time: 2 HOURS | Finishing time: 10 MINUTES | 139°F (59.4°C)

Pork tenderloin is the tenderest part of a pig and, like beef tenderloin, is a very small cut compared to other parts of the animal. As such, it can be more expensive than chops or ribs, but it results in a particularly flavorful pork dish. The best way to lock in the flavor and add a flavorful crust is to use a dry rub versus a marinade. This is great served on a bed of Garlic and Herb Cheese Cauliflower Purée (page 56) and with a side of Garlic-Parmesan Asparagus (page 57). **6 TO 8 SERVINGS**

1 tablespoon garlic powder

1 tablespoon ground cumin

1 tablespoon dried thyme

1 tablespoon dried rosemary

Pinch lime sea salt

1 (3-pound) pork tenderloin, silverskin removed

1 tablespoon extra-virgin olive oil

1 tablespoon kosher salt

2 tablespoons unsalted butter

1. Set your sous vide circulator to 139°F (59.4°C).

2. Combine the garlic powder, cumin, thyme, rosemary, and lime salt in a small bowl and mix thoroughly.

3. Place the pork on a clean flat surface, brush it first with the olive oil, then rub with the kosher salt, followed by the garlic-herb rub.

4. Transfer the pork to a large food-safe bag (or use a sharp knife to cut the tenderloin evenly into two pieces, and place in two medium food-safe bags).

5. Seal using your vacuum sealer or using the water displacement method. Place the sealed bag(s) in the water bath and cook for 2 hours.

6. When the pork is cooked, remove it from the bag, place it on a paper towel–lined plate or baking sheet, and pat it dry. Discard the cooking liquid.

7. Heat a cast iron pan over high heat and add the butter. Add the pork to the pan and sear until browned on all sides, 3 to 4 minutes in total.

8. Remove the pork from the pan and transfer it to a cutting board. Let it rest for 5 minutes before carving into 1- to 2-inch medallions.

CHICKEN CORDON BLEU

Prep time: 15 MINUTES, PLUS 2 TO 3 HOURS COOLING TIME | Cook time: 1 HOUR, 30 MINUTES
Finishing time: 5 MINUTES | 140°F (60°C)

I've seen several versions of stuffed chicken, but the most famous, and arguably the most delicious, is the French classic chicken cordon bleu. Translated, cordon bleu simply means blue ribbon and it signified members of the highest order of French knighthood. It is also the largest culinary education institution in the world. This simple dish requires a bit of prep and practice to ensure you don't lose all the stuffing, but it's worth it when the melted cheese runs out on to your plate. **4 SERVINGS**

4 boneless, skinless chicken breasts, tenders removed and saved for another purpose
2 tablespoons sea salt, divided
6 teaspoons freshly ground black pepper, divided

1 teaspoon smoked paprika
½ cup (8 tablespoons) unsalted butter, divided
4 garlic cloves, pressed
8 slices Black Forest ham
8 slices Swiss or Gruyère cheese

1. Set your sous vide circulator to 140°F (60°C).

2. Place the chicken breasts on a cutting board and sprinkle with 2 teaspoons of salt and 1 teaspoon of pepper. Cover the breasts with plastic wrap and use a rolling pin to flatten them as much as possible. They should expand in width but decrease in thickness.

3. Heat the remaining 5 teaspoons of pepper, paprika, ¼ cup of butter, and garlic in a small saucepan over medium heat until the butter has completely melted. Remove from the heat and transfer to a small bowl.

4. Uncover the breasts and brush on one side with the butter mixture. Lay 2 slices of ham and 2 slices of cheese on each breast, covering it.

5. On a clean surface, lay a new piece of plastic wrap for each breast. Place the chicken in the middle of the wrap and roll it as tightly as possible, sealing the wrap around the cutlet. Repeat this for a total of four rolled breasts. Refrigerate for 2 to 3 hours or in the freezer for 20 to 30 minutes to allow the chicken to firm up.

6. Transfer two breasts into each of two food-safe bags and seal using your vacuum sealer or the water displacement method. Place the sealed bags in the water bath and cook for 1 hour, 30 minutes.

7. When the breasts are cooked, remove them from the bag and discard the plastic wrap.

8. Melt the remaining ¼ cup of butter in a cast iron pan over medium-high heat and sear the chicken for 1 to 2 minutes per side, finishing the edges with a blowtorch (if one is available). Serve promptly!

CRISPY FRIED CHICKEN DRUMSTICKS

Prep time: 5 MINUTES | Cook time: 3 HOURS | Finishing time: 15 MINUTES | **155°F (68°C)**

Crispy skin! Mouthwatering crunch! Juicy, perfectly cooked chicken. That's what you get with this recipe. Cooking drumsticks sous vide takes the guesswork out of the blind cooking traditional method of frying chicken. Because we cook the chicken throughout before it hits the hot oil, you're able to control how much crisp and fry you want without having any concerns about how well cooked the chicken will be. We recommend typical Southern fare to pair with this recipe such as macaroni salad, waffles, fresh slaw, or perhaps a healthier sous vide Lemony Garlic Kale Salad (page 65). **8 DRUMSTICKS**

8 chicken drumsticks (about 2 pounds)

6 teaspoons kosher salt, divided

2 gallons canola oil, for frying

3 cups all-purpose flour

2 tablespoons garlic powder

2 tablespoons freshly ground
white pepper

1½ tablespoons chipotle chili powder

1 tablespoon dried mustard

1 tablespoon freshly ground black pepper

½ tablespoon dried basil

2¼ cups buttermilk

1. Set your sous vide circulator to 155°F (68°C).

2. Season the drumsticks all over with 3 teaspoons of salt, then place them in a food-safe bag in a single layer. Use more than one bag if necessary, to ensure a single layer. Seal the bag(s) using your vacuum sealer or the water displacement method. Place the sealed bag in the water bath and cook for 3 hours.

3. Meanwhile, heat the oil in a large stockpot to between 375°F and 400°F (190°C to 204°C). (You can measure the temperature with a candy thermometer.) Set the oven temperature to the lowest temperature setting. Place a wire rack over a baking sheet and place it in the oven.

4. Whisk together the remaining 3 teaspoons of salt, flour, garlic powder, white pepper, chipotle chili powder, mustard, black pepper, and basil in a large mixing bowl.

5. When the chicken is cooked, remove the drumsticks from the bag, place them on a paper towel–lined plate or baking sheet, and pat them dry. Discard the cooking liquid.

6. Pour the buttermilk into a large mixing bowl and place next to the flour mixture. Taking one drumstick at a time, roll it in the flour mix, dip it in the buttermilk, and roll it in the flour mix again. Place the breaded chicken on a separate baking sheet.

7. Carefully slide 2 or 3 drumsticks into the hot oil and fry until the crust is golden and crispy, 3 to 4 minutes. Remove the chicken, letting any excess oil drip off before placing the drumsticks on the wire rack in the oven to keep warm. Repeat with the remaining drumsticks.

SWEET CHILI CHICKEN LETTUCE WRAPS

Prep time: 10 MINUTES | Cook time: 1 HOUR, 30 MINUTES | Finishing time: 5 MINUTES | 140°F (60°C)

This interactive dish allows your guests to build their appetite while building their own meal. Made famous at the legendary P.F. Chang's restaurant, it's an easy dish to re-create and it's fun for your friends and family to assemble their wraps however they choose. The crispness of the iceberg lettuce gives additional texture and freshness to counter the sweetness of the hoisin sauce. **4 TO 6 SERVINGS**

½ cup sweet chili sauce

3 tablespoons soy sauce

2 tablespoons freshly grated ginger

1 tablespoon honey or raw sugar

2 garlic cloves, minced

Juice of 1 lime, divided

4 chicken breasts

¼ tablespoon kosher salt

¼ tablespoon freshly ground
  black pepper

½ cup hoisin sauce

1 head iceberg lettuce

⅛ cup sesame seeds

4 scallions, diced

1. Set your sous vide circulator to 140°F (60°C).

2. Mix together the chili sauce, soy sauce, ginger, honey, garlic, and half the lime juice in a small bowl. Heat the mixture in a small saucepan over medium heat for 5 minutes.

3. Season the chicken breasts with the salt and pepper. Put two chicken breasts in a single layer in each of two food-safe bags, and divide the sweet chili sauce evenly between the bags.

4. Seal using your vacuum sealer or using the water displacement method, ensuring the chicken remains in a single layer. Place the sealed bags in the water bath and cook for 1 hour, 30 minutes.

5. When the chicken is cooked, remove the breasts from the bag and pat them dry. Discard the cooking liquid. Using a sharp knife, cut the breasts into 1-inch cubes.

6. Put the hoisin sauce in a medium bowl, add the chicken, and toss thoroughly, making sure the chicken pieces are thoroughly coated.

7. Pull 12 to 16 large leaves from the lettuce. Rinse each piece and pat dry with paper towels. Stack in piles of 6 or 8.

8. Transfer the chicken pieces to a serving dish. Add chicken to a lettuce leaf and top with the sesame seeds and scallions before wrapping.

MAUI RIBS

Prep time: 20 MINUTES | Cook time: 16 HOURS | Finishing time: 30 MINUTES | 160°F (71.1°C)

Hawaii is one of my favorite places to visit. I have such fond memories of vacationing at Kaanapali Beach in Maui with my family. On one of our trips, we attended a luau and got to try Kalua pig. The pig is cooked for hours in a six-foot-deep pit on the beach, surrounded by hot rocks and banana leaves and covered in sand. Cooked this way, the pork retains its moisture. In a way, sous vide cooking creates similar results—low cooking at a lower temperature to retain moisture and inject flavor—without your backyard getting dug up! **4 TO 6 SERVINGS**

4 pounds (2 racks) baby back ribs, membrane removed

3 cups Maui Pork Marinade (page 36), ½ cup reserved

⅓ cup ketchup

4 scallions, chopped finely

1. Set your sous vide circulator to 160°F (71.1°C).

2. Put the ribs in a large food-safe bag or use a sharp knife to separate the ribs into four smaller portions. Pour the 3 cups of marinade evenly among the bags and seal using your vacuum sealer or the water displacement method. Place the sealed bag in the water bath and cook for 16 hours. Cover the container with a lid, plastic wrap, or another method to limit water evaporation.

3. Meanwhile, mix together the remaining ½ cup of marinade and ketchup in a medium bowl. Refrigerate if prepared before the ribs are cooked.

4. When the ribs are cooked, remove them from the bag, put them on a baking sheet, and pat them dry. Discard the cooking liquid.

continued

5. Preheat the oven to 300°F (150°C).

6. Using a soft brush, brush sauce on both sides of the ribs and place in the oven. Cook for 10 minutes, then remove the ribs from the oven and brush on a second coat of sauce. Continue for 30 minutes total until a sticky bark has formed.

7. Cut the ribs into individual pieces, garnish with the scallions, and serve.

KARL'S JERK CHICKEN

Prep time: 20 MINUTES, PLUS 12 HOURS MARINATING TIME | Cook time: 2 HOURS
Finishing time: 10 MINUTES | **150°F (65.5°C)**

*When my wife and I moved to London from Toronto, it was hard to leave every-
thing behind—my family, friends, and coworkers, as well as my precious trio
of sous vide machines (the UK uses different voltages than North America). I
decided to share my babies with some special friends who I knew would put them
to good use and take great care of them. Karl, a friend from university and an avid
home cook, didn't waste any time diving right in. One of his first dishes was jerk
chicken and it was delicious, so we decided to share it.* **6 TO 8 SERVINGS**

4 to 6 whole Scotch bonnet peppers
6 scallions, chopped
6 garlic cloves, pressed
1 thumb fresh ginger, chopped
½ cup soy sauce
¼ cup extra-virgin olive oil
¼ cup freshly squeezed lime juice
2 tablespoons freshly grated lime zest

2 tablespoons brown sugar
2 tablespoons fresh thyme leaves
1 tablespoon ground allspice
1 tablespoon kosher salt
2 teaspoons freshly ground black pepper
1 teaspoon ground nutmeg
8 to 12 chicken thighs or legs

1. Combine the peppers, scallions, garlic, ginger, soy sauce, olive oil, lime juice,
lime zest, brown sugar, thyme, allspice, salt, pepper, and nutmeg in a food processor and blend. Reserve ¼ cup of sauce for basting.

2. Place the chicken thighs and jerk sauce in a food-safe bag and seal using
your vacuum sealer or the water displacement method. Place the sealed bag in
the refrigerator and let marinate for 12 hours.

3. Set your sous vide circulator to 150°F (65.5°C). Place the sealed bag in the
water bath and cook the chicken for 2 hours.

4. When the chicken is cooked, transfer it to a paper towel–lined plate or baking
sheet and pat it dry. Discard the cooking liquid.

5. Brush the chicken with the reserved fresh jerk sauce.

6. Heat a cast iron pan over high heat, or heat your grill to 400°F, and sear the
chicken for 30 seconds per side.

BUTTER CHICKEN

Prep time: 55 MINUTES | Cook time: 1 HOUR, 30 MINUTES | Finishing time: 10 MINUTES | **140°F (60°C)**

Unlike other Indian curries, the flavors here are mild and the sauce creamy. The slightly longer cooking time needed for sous vide versus traditional recipes only enhances the deep, rich flavor. **4 SERVINGS**

4 boneless skinless chicken breasts
½ teaspoon salt

½ teaspoon freshly ground black pepper
5 cups Butter Chicken Sauce (page 33)

1. Set your sous vide circulator to 140°F (60°C).

2. Season the chicken breasts on both sides with salt and pepper. Put two chicken breasts in each of two food-safe bags, divide the sauce evenly between the bags, and seal using your vacuum sealer or the water displacement method. Place the sealed bags in the water bath and cook for 1 hour, 30 minutes.

3. When the chicken is cooked, remove the breasts from the bags, transfer to a cutting board, and pat them dry.

4. Pour the cooking liquid from the bags into a large saucepan and cook over medium-high heat, stirring often, allowing the mixture to reduce for 10 minutes.

5. Meanwhile, use a sharp knife to cut the chicken into 2-inch chunks.

6. Add the chicken to the sauce and stir for 2 to 3 minutes to heat through. Transfer to a serving dish and serve with warm naan bread for dipping!

PROSCIUTTO-WRAPPED TURKEY BREAST STUFFED *with* LEMON *and* SAGE

Prep time: 30 MINUTES | Cook time: 3 HOURS | Finishing time: 15 MINUTES | **145°F (62.7°C)**

Last Thanksgiving, my wife and I bravely decided to flip the turkey tradition on its head. Instead of roasting a whole turkey as my mom and sisters had always done, we decided to try turkey sous vide. Now, cooking an entire turkey sous vide is possible, but challenging, plus you won't end up with that golden-brown roasted flavor. We instead elected to cook a few large stuffed turkey breasts, wrapping them in thinly sliced ham. My nieces and nephews, often my toughest critics, were blown away, and I fear the roasted turkey of old will become a faint memory. Serve these turkey medallions with Garlic and Herb Cheese Cauliflower Purée (page 56), fresh gravy, Butternut Squash Soup (page 59), or Crispy-Skin Smashed Potatoes (page 49). **4 TO 6 SERVINGS**

1 white onion, chopped

3 tablespoons unsalted butter, divided

1 cup freshly cut pancetta cubes

4 tablespoons pine nuts

2 tablespoons chopped fresh sage

4 garlic cloves, pressed

Zest of 2 lemons

4 tablespoons chopped fresh parsley

¾ cup white or whole-wheat bread crumbs

1 egg, beaten

1 (4-pound) boneless turkey breast, butterflied, skin removed and reserved

½ teaspoon salt

½ teaspoon freshly ground black pepper

16 slices fresh prosciutto

1. Set your sous vide circulator to 145°F (62.7°C).

2. Sauté the onion in 2 tablespoons of butter in a large skillet over medium heat until softened, about 10 minutes. Transfer to a medium bowl.

3. Add the pancetta to the skillet and brown for 5 minutes over medium heat. Add the pine nuts, sage, garlic, and lemon zest and cook for 2 minutes until the pine nuts begin to brown. Add the parsley and stir.

4. Add the pancetta–pine nuts mix to the onion. Add the bread crumbs and egg and stir to mix.

continued

5. Lay the turkey breast skin flat on a parchment paper–lined baking sheet. Season with the salt and pepper, then lay another piece of parchment paper on top and press flat with a second baking sheet. Leave the baking sheet on top and place in the refrigerator.

6. Cover the skinless turkey breast with wax paper or plastic wrap and use a rolling pin or meat hammer to pound it to about a 2-inch thickness. This may take some time. Work your way from the middle to the outside and ensure the breast is evenly flat and has a rectangular shape (trim the edges if necessary).

7. Lay the prosciutto in the middle of a large aluminum foil–covered and greased baking sheet so the slices overlap by about 1 inch. Place the flattened turkey breast on the prosciutto grid and spoon the stuffing down the center of the turkey breast, creating a strip of stuffing that nearly touches the edges of the breast.

8. Roll the turkey breast as tightly as possible from one side to the other. Continue to hold it tightly while folding the prosciutto around the rolled breast, alternating sides, until the turkey is completely wrapped.

9. Lay a large square piece of plastic wrap on the countertop. Transfer the rolled and wrapped breast to the plastic wrap and use the wrap to tightly hold the turkey together. You should be able to easily move the breast around when it is tightly wrapped.

10. Put the breast in a large food-safe bag and seal using your vacuum sealer or using the water displacement method. Place the sealed bag in the water bath and cook for 3 hours.

11. Thirty minutes before the end of the cooking time, preheat the oven to 400°F (205°C).

12. Remove the flattened skin from the refrigerator, keeping it pressed between the two baking sheets, and place on the middle rack of the oven. Cook for 30 minutes, then remove from the oven, remove the top baking sheet and the top layer of parchment paper, and allow the skin to cool to room temperature.

13. When the turkey breast is cooked, carefully remove the plastic wrap and discard.

14. Melt the remaining 1 tablespoon of butter in a large skillet over medium-high heat. Add the turkey breast to the skillet and let the prosciutto sear for about 45 seconds per side, rolling the turkey to ensure all sides are browned, for a total of 2 to 2½ minutes.

15. Transfer the breast to a large cutting board and use a very sharp knife to slice it into 1½- to 2-inch medallions.

16. Use a rolling pin or other blunt object to smash the now-crisp turkey skin to break it into shards, add them to a large serving bowl, and encourage your guests to add it to their plates!

TURKEY BURGERS

Prep time: 15 MINUTES | Cook time: 1 HOUR, 15 MINUTES | Finishing time: 10 MINUTES | **147°F (63.8°C)**

A summer classic, the grilled burger is a lunch and dinnertime favorite. For those of you looking for a healthier option than a beef burger, use your sous vide to unlock the juicy flavors of ground turkey. Ground turkey has less fat and fewer calories than its beef counterpart and turkey burgers have a reputation for being drier than traditional beef burgers, but when you cook them sous vide they'll be juicy and delicious! **6 BURGERS**

1½ pounds lean ground turkey
16 cream crackers, crushed
2½ tablespoons finely chopped
 fresh parsley
2 tablespoons finely chopped fresh basil
½ tablespoon Worcestershire sauce
½ teaspoon garlic powder

1 large egg
6 teaspoons extra-virgin olive oil
6 white or whole-wheat buns, toasted
6 tomato slices
6 large pieces iceberg lettuce
6 slices Cheddar cheese (optional)

1. Set your sous vide circulator to 147°F (63.8°C).

2. Combine the turkey, crackers, parsley, basil, Worcestershire sauce, and garlic powder in a large bowl. Crack the egg into the bowl and use your hands to thoroughly mix.

3. Line a baking sheet with parchment or wax paper and divide the mix into 6 patties. If you have a burger press, use it! Cover the patties with plastic wrap and place in the refrigerator for 2 hours to allow the meat to bind and firm up.

4. Remove from the refrigerator and quickly place up to two burgers into each of three food-safe bags. Seal using your vacuum sealer or the water displacement method. Place the sealed bags in the water bath and cook for 1 hour, 15 minutes.

5. When the patties are cooked, remove them from the bag and discard any cooking liquids.

6. Heat the olive oil in a large cast iron pan over high heat. Gently place the burgers in the pan and cook for 45 seconds on each side.

7. Serve on a freshly toasted bun with tomato, lettuce, and cheese (if using). They go well with Sweet Potato Fries with Malt Vinegar Mayonnaise (page 50) as a side!

PIRI PIRI CAPON WINGS

Prep time: 45 MINUTES | Cook time: 2 HOURS | Finishing time: 15 MINUTES | 147°F (63.8°C)

A few years ago, some friends and I were out and witnessed some extra-meaty wings pass us on a server's tray. The wings were nowhere on the menu. The server told us they were capon wings, a secret item. A few minutes later, our secret wings arrived and they were huge! Not just in size, but their meatiness was different, too. Capons are castrated cockerels specifically fattened for eating; castration improves the quality and flavor of the flesh. Capon wings aren't easy to find, so go ahead and use regular chicken wings; they'll still taste great. **2 TO 4 SERVINGS**

3 pounds capon chicken wings   2½ cups Piri Piri Sauce (page 31)

1. Set your sous vide circulator to 147°F (63.8°C).

2. Season the capon wings with salt and pepper. Place them in a single layer in a food-safe bag. Pour the piri piri sauce evenly across the wings, reserving ¼ cup, and seal using your vacuum sealer or the water displacement method. Place the sealed bag in the water bath and cook for 2 hours.

3. Set the oven to broil.

4. When the wings are cooked, transfer them to a large bowl. Pour the reserved piri piri sauce over the wings and toss. Using tongs, place the wings on an aluminum foil–lined baking sheet. Pour any remaining sauce over the wings and transfer to the oven.

5. Broil for 10 minutes, flipping them after 5 minutes. Remove from the oven and serve.

CARAMELIZED FOIE GRAS

Prep time: 10 MINUTES | Cook time: 30 MINUTES | Finishing time: 5 MINUTES | 136°F (57.7°C)

This dish goes back to the roots of sous vide cooking, allowing you to cook a very rich and relatively expensive delicacy without fear of ruining your investment. In 1974, Georges Pralus, chef of the restaurant Troisgros in Roanne, France, discovered that when cooking foie gras using sous vide methods instead of traditional roasting, the duck liver kept its original shape, retained most of its fat, and had better texture. While sous vide cooking methods are becoming increasingly popular today, it was taboo in restaurants, even into the 1990s. It was viewed as only applicable for industrial food manufacturing, so much so that when the famous French restaurant Chez Margot was discovered using sous vide methods, it was forced into bankruptcy! **6 TO 8 SERVINGS**

1 pound duck liver lobes
Sea salt

1 fresh baguette or brioche,
 thinly sliced and toasted

1. Set your sous vide circulator to 136°F (57.7°C).

2. Separate the two lobes of the liver and cut into 8 to 12 (1-inch) slices. (They should be in the shape of a semicircle with a small groove down the middle.)

3. Add the desired number of slices, in a single layer and not touching the others, to a food-safe bag and seal using the water displacement method. (Foie gras is very delicate and some vacuum sealers may crush the shape of the liver. If you have a vacuum sealer that has manual controls, you can manually set your pressure and limit the squeeze). Place the sealed bag in the water bath and cook for 30 minutes.

4. When the foie gras is cooked, remove the slices and place them on paper towels to drain the extracted fat.

5. Place the slices on a fireproof surface and use a blowtorch to caramelize the liver to give it a smoked flavor similar to that of roasted foie gras.

6. Sprinkle with salt and serve the slices with thinly sliced, toasted bread.

PORK BELLY TACOS

Prep time: 5 MINUTES | Cook time: 7 HOURS | Finishing time: 5 MINUTES | **176°F (80°C)**

Pork belly has been gaining popularity these past few years. It can be tricky to cook in the oven, and if it's not done correctly, it can be super fatty, which turns some people off. Pork belly is one of the easiest things to make sous vide. It's on the longer side of cooking times, but if you prepare it in the morning, you'll have the perfect base for delicious dinner tacos! **8 TACOS**

½ cup freshly squeezed orange juice
4 tablespoons honey or raw
  sugar, divided
2 tablespoons pressed fresh garlic
2 tablespoons pressed fresh ginger
2 tablespoons soy sauce
2 teaspoons hoisin sauce

2 teaspoons sriracha
Zest of ½ orange
1 pound pork belly
8 flour tortillas, warmed
½ cup chopped fresh cilantro
½ lime, cut into wedges

1.   Set your sous vide circulator to 176°F (80°C).

2.   Mix together the orange juice, 3 tablespoons of honey, garlic, ginger, soy sauce, hoisin, sriracha, and orange zest in a medium bowl.

3.   Place the pork belly in a large food-safe bag. Pour the marinade into the bag, covering the belly, then seal using your vacuum sealer or the water displacement method. Place the sealed bag in the water bath and cook for 7 hours.

4.   When the belly is cooked, transfer it to a baking sheet.

5.   Transfer the cooking liquid to a medium saucepan over high heat with the remaining 1 tablespoon of honey and cook, stirring often, until bubbling and reduced by half, about 5 minutes.

6.   Remove from the heat and brush all sides of the belly with the glaze. Reserve any remaining glaze to use as sauce.

7.   Using a sharp knife, cut the pork into 1-inch slices.

8.   Place the belly in tortillas, sprinkle with the cilantro, squeeze lime juice over, and drizzle with the reserved sauce.

WHILE YOU WERE SLEEPING SOUS VIDE BACON

Prep time: 1 MINUTE | Cook time: 8 HOURS | Finishing time: 5 MINUTES | **145°F (62.7°C)**

Yes, you can cook bacon sous vide. A lot of people ask why would you want to wait 8 hours for bacon. Well, you wouldn't. What you might do, however, is cook ultra-tender bacon while you're sleeping. How much cooking have you done in your sleep before? Probably none! There isn't much to this recipe, but what you end up with is very tender bacon, far more tender than you can get in a pan. **1 POUND BACON, 4 SERVINGS**

1 (1-pound) package bacon

1. Set your sous vide circulator to 145°F (62.7°C).

2. Place the bacon in its original sealed package into the water bath and cook for at least 8 hours.

3. Open the package and remove the bacon slices.

4. Heat a large cast iron pan over medium-high heat and add the bacon in a single layer. Cook for 2 to 3 minutes until the bacon begins to crisp.

5. Remove the bacon from the pan, place on a paper towel–lined plate or baking sheet to remove excess fat, and serve.

BUFFALO CHICKEN BREASTS

Prep time: 5 MINUTES | Cook time: 1 HOUR 30 MINUTES | Finishing time: 5 MINUTES | 140°F (60°C)

Arguably one of my favorite flavors, this is a spicy take on the classic sports bar chicken wing flavor. Simple to make and easy to control the heat, this dish is versatile and can be served over rice or a fresh crisp salad. **4 SERVINGS**

½ **cup hot sauce**
2 **tablespoons unsalted butter, divided**
1 **tablespoon white vinegar**

4 (4-ounce) chicken breast halves
½ teaspoon kosher salt
½ teaspoon freshly ground black pepper

1. Set your sous vide circulator to 140°F (60°C).

2. Combine the hot sauce, 1 tablespoon of butter, and vinegar in a small sauce-pan over medium heat, and heat until the butter melts. Set aside.

3. Season the chicken breasts with the salt and pepper.

4. Place two chicken breasts in each of two food-safe bags, divide the buffalo sauce evenly between the bags, and seal using your vacuum sealer or the water displacement method. Add the sealed bags to the water bath and cook for 1 hour, 30 minutes.

5. When the chicken is cooked, remove the breasts from the bag, transfer to a paper towel–lined plate or baking sheet, and discard the cooking liquid.

6. Heat the remaining 1 tablespoon of butter in a cast iron pan over high heat until it begins to bubble. Add the chicken breasts and cook until they begin to brown, about 1 minute per side.

7. Remove from the pan and slice into 1-inch strips.

8. Serve with rice or in a fresh chopped salad.

CREAMY LEMON CHICKEN

Prep time: 10 MINUTES | Cook time: 1 HOUR, 30 MINUTES | Finishing time: 10 MINUTES | 140°F (60°C)

This citrus chicken dish won't leave a sour taste in your mouth! The addition of cream, garlic, and broth helps cut the acidity and allows the fresh lemon flavors to penetrate the chicken, leaving you with a lemony fresh taste, whether it is served hot or cold. **4 SERVINGS**

4 boneless skinless chicken breasts
½ teaspoon kosher salt
½ teaspoon freshly ground black pepper
Zest of 1 lemon
¼ cup freshly squeezed lemon juice
¼ cup heavy (whipping) cream
3 tablespoons unsalted butter, divided

2 tablespoons chicken broth
1 tablespoon chopped fresh thyme leaves
1 tablespoon extra-virgin olive oil
3 garlic cloves, pressed
1 small shallot, chopped
1 large lemon, thinly sliced

1.   Set your sous vide circulator to 140°F (60°C).

2.   Place the chicken breasts on a clean flat surface and season with the salt and pepper.

3.   Combine the lemon zest and juice, cream, 2 tablespoons of butter, broth, thyme, olive oil, garlic, and shallot in a small saucepan over medium heat and heat, stirring often, until the butter has melted.

4.   Transfer two chicken breasts into each of two food-safe bags and divide the lemon butter evenly between the bags. Place 1 lemon slice on each breast and seal using your vacuum sealer or the water displacement method. Add the sealed bags to the water bath and cook for 1 hour, 30 minutes.

5.   When the chicken is cooked, remove the breasts from the bag, transfer to a paper towel–lined plate or baking sheet, and discard the cooking liquid.

6.   Heat the remaining 1 tablespoon of butter in a cast iron pan over high heat until it begins to bubble, then add the chicken breasts and cook for 1 minute per side until they begin to brown.

7.   Remove from the pan and slice into 1-inch strips.

8.   Serve with rice or in a fresh chopped salad.

PEANUT CHICKEN SATAY

Prep time: 15 MINUTES | Cook time: 1 HOUR, 30 MINUTES | Finishing time: 10 MINUTES | **140°F (60°C)**

A common staple dish in Thailand, these peanut chicken skewers bring in the nutty flavors of peanuts combined with some umami hints and the sweetness of honey and garlic. These are great as a starter or a main course; just remember to reserve some of the marinade as a sauce for dipping! **4 SERVINGS**

4 boneless skinless chicken breasts
½ teaspoon kosher salt
½ teaspoon freshly ground black pepper

½ cup Peanut Chicken Marinade (page 38)
2 tablespoons unsalted butter

1. Set your sous vide circulator to 140°F (60°C). If using wooden skewers, fill a large bowl with water and soak them while the chicken cooks.

2. Season the chicken breasts with the salt and pepper.

3. Place two chicken breasts into each of two food-safe bags, divide the peanut sauce evenly over the breasts, ensuring they are covered, and seal using your vacuum sealer or the water displacement method. Add the sealed bags to the water bath and cook for 1 hour, 30 minutes.

4. When the chicken is cooked, remove the breasts from the bag, transfer to a paper towel–lined plate or baking sheet, and discard the cooking liquid.

5. Heat the butter in a cast iron pan over high heat until it begins to bubble. Add the chicken breasts and cook until they begin to brown, about 1 minute per side.

6. Remove from the pan and slice into 1-inch pieces. Thread 4 or 5 pieces on each skewer, and serve.

TEQUILA-LIME CHICKEN BREASTS

Prep time: 15 MINUTES | Cook time: 1 HOUR, 30 MINUTES | Finishing time: 10 MINUTES | **140°F (60°C)**

This might as well have been called margarita chicken! It has all the fixings for a summer cocktail, and it transforms the chicken into a fresh, citrusy, creamy combination. It's rich in flavor and is perfect on a hot summer night! **4 SERVINGS**

4 boneless skinless chicken breasts
¼ teaspoon salt
¼ teaspoon black pepper

1½ cups Tequila Lime Sauce (page 30)
8 tablespoons unsalted butter, cubed

1. Set your sous vide circulator to 140°F (60°C).

2. Season the chicken breasts with the salt and pepper. Place two chicken breasts into each of two food-safe bags, divide the tequila-lime sauce evenly over the breasts ensuring they are covered, and seal using your vacuum sealer or the water displacement method. Add the sealed bags to the water bath and cook for 1 hour, 30 minutes.

3. When the chicken is cooked, remove the breasts from the bag, transfer to a paper towel–lined plate or baking sheet, and discard the cooking liquid.

4. Heat the butter in a cast iron pan over high heat until it begins to bubble. Add the chicken breasts and cook until they begin to brown, about 1 minute per side.

5. Remove from the heat, slice into 1-inch strips, and serve.

chapter six
Red Meats

Sous vide steak was what got me hooked on sous vide cooking. From tender cuts like beef tenderloin and rib eye steaks through to the toughest cuts like brisket and chuck roasts, you can cook them all into mouthwatering, tenderized, and flavor-packed dishes. The recipes in this chapter are focused on beef and lamb with a few dishes you might not eat very often, like veal, goat, venison, and even alpaca. Red meats are enjoyed all around the world and I've included dishes from the United States, India, Jamaica, Cuba, South Africa, Peru, and Korea.

ROSEMARY-GARLIC RACK OF LAMB (page 144)

CITRUS-GARLIC CHUCK ROAST

Prep time: 20 MINUTES | Cook time: 24 HOURS | Finishing time: 10 MINUTES | **140°F (60°C)**

One of the longest, most rewarding cooks you can do, chuck roast transforms beautifully after a full day in the water bath. The long cooking time breaks down the connective tissues and tenderizes the meat into an amazingly flavorful (and affordable) roast. **8 TO 10 SERVINGS**

2 pounds beef chuck, up to 3 inches thick

1 teaspoon kosher salt

1 teaspoon freshly ground black pepper

4 tablespoons extra-virgin olive oil, divided

1 cup soy sauce

½ cup freshly squeezed lemon juice

½ cup freshly squeezed orange juice

½ cup Worcestershire sauce

¼ cup Dijon mustard

3 garlic cloves, minced

1. Set your sous vide circulator to 140°F (60°C).

2. Place the beef on a clean flat surface and season with the salt and pepper. Using butcher's twine, truss the roast with the string, spacing out sections 2 to 3 inches apart.

3. Heat a cast iron pan over high heat. While it heats up, use a soft brush to baste the beef all over with 2 tablespoons of olive oil. Gently add the beef to the pan and sear for 1 minute on each side. Remove from the pan.

4. Mix together the soy sauce, lemon juice, orange juice, Worcestershire sauce, mustard, and garlic in a medium bowl.

5. Place the beef in a food-safe bag, pour the marinade into the bag, and seal using your vacuum sealer or the water displacement method. Add the sealed bag to the water bath and cook for 24 hours.

6. When the beef is cooked, remove the roast and pour the cooking liquids into a small saucepan. Heat on high until reduced by half, about 10 minutes.

7. Into a cast iron pan over high heat, pour the remaining 2 tablespoons of olive oil. Carefully place the beef in the pan, searing for 1 minute on all sides.

8. Remove the roast from the pan and let it rest for 5 minutes.

9. Slice the beef to your desired thickness, plate, and pour the reduced jus (sauce) on top.

BARBARIAN RIB STEAK

Prep time: 5 MINUTES | Cook time: 90 MINUTES | Finishing time: 5 MINUTES | 133°F (56.1°C)

This is a satisfying purchase, if you're able to find one of these cuts at your local butcher. It looks like something a caveman might enjoy, but it's got considerably more class. Technically it's just a bone-in rib eye, but the long protruding bone resembles a tomahawk, which is where it gets its much cooler name. This cut of beef is likely too large for one person, but perfect for sharing if you've got some good sides to serve it with. **2 SERVINGS**

1 pound barbarian rib steak or tomahawk rib steak

¼ teaspoon kosher salt

¼ teaspoon freshly ground black pepper

½ teaspoon garlic powder

1 tablespoon butter

1. Set your sous vide circulator to 133°F (56.1°C). This will give you just over medium rare, plus a few degrees to account for the bone.

2. Rub the meat with the salt, pepper, and garlic powder. Place the beef in a food-safe bag, add the butter to the bag, and seal using your vacuum sealer or the water displacement method. Add the sealed bag to the water bath and cook for 90 minutes.

3. When the beef is cooked, remove it from the bag and pat it dry. Discard the cooking liquid.

4. Preheat the grill to between 600°F and 800°F—the hotter the better—or heat a cast iron pan over high heat.

5. Place the steak on the hot grill or in the pan and sear for 1 minute per side.

6. Remove from the grill or pan and let rest for 5 minutes before carving.

CHATEAUBRIAND

Prep time: 15 MINUTES | Cook time: 2 HOURS | Finishing time: 10 MINUTES | 133°F (56.1°C)

Chateaubriand doesn't need a long cook to tenderize the beef since it is part of the tenderloin. The center cut is the thickest portion of the tenderloin, and after 2 hours in the water bath, it can almost fall apart (use the butcher's twine just in case). Due to the lack of fat in the tenderloin, its natural flavors are somewhat faint, so be sure to use the correct amount of salt and aromatics to help inject some flavor. **4 TO 6 SERVINGS**

1 (2¼-pound) center-cut beef tenderloin
½ teaspoon kosher salt
½ teaspoon freshly ground black pepper
4 tablespoons unsalted butter, divided

1 shallot, chopped
2 fresh thyme sprigs
1 fresh rosemary sprig

1. Set your sous vide circulator to 133°F (56.1°C).

2. Lay the tenderloin on a clean flat surface and season all over with the salt and pepper. Using butcher's twine, truss the tenderloin, creating 3 to 4 sections.

3. Heat 2 tablespoons of butter in a large cast iron pan over high heat. When the butter begins to bubble, gently place the trussed tenderloin in the pan and sear each side for 30 to 45 seconds.

4. Remove the tenderloin and add the shallot, thyme, and rosemary to the pan. Spoon the butter over the herbs and cook until they become bright green and soften, 1 to 2 minutes.

5. Add the trussed tenderloin to a food-safe bag, pour the butter-herb mix into the bag, and seal using your vacuum sealer or the water displacement method. Add the sealed bag to the water bath and cook for 2 hours.

6. When the tenderloin is cooked, remove it from the bag and place on a paper towel–lined plate or baking sheet. Discard the cooking liquid.

7. Melt the remaining 2 tablespoons of butter in a cast iron pan over high heat. Add the tenderloin when the butter begins to bubble and smoke. Sear for 2 minutes on all sides.

8. Remove the tenderloin from the heat and let it stand for 5 minutes before carving into 2- to 3-inch medallions. Serve with vegetables and potatoes.

COFFEE-CRUSTED NEW YORK STRIP LOIN

Prep time: 15 MINUTES | Cook time: 1 HOUR, 30 MINUTES | Finishing time: 10 MINUTES | 131°F (55°C)

Coffee and steak—who knew? I first experienced this flavor combination when using a premade rub by a popular barbecue manufacturer. While New York strip loins already have a great flavor when simply seasoned with salt and pepper, there is always room for improvement. The combination of sugar, garlic, paprika, and coffee might seem odd, but they add an excellent flavor mix and help build a unique crust upon searing. Serve this loin with Crispy-Skin Smashed Potatoes (page 49) and Garlic-Parmesan Asparagus (page 57). **4 SERVINGS**

1 tablespoon brown sugar

1 tablespoon kosher salt

1 tablespoon freshly ground black pepper

1 tablespoon coffee grounds

1 tablespoon garlic powder

1 tablespoon smoked paprika

4 (2-inch-thick) New York strip loins

1 tablespoon extra-virgin olive oil

2 tablespoons unsalted butter

1.   Set your sous vide circulator to 131°F (55°C).

2.   Mix together the brown sugar, salt, pepper, coffee grounds, garlic powder, and paprika in a small bowl.

3.   Lay the steaks on a clean flat surface and use a soft brush to coat the steaks with a thin layer of olive oil. Shake the rub evenly on both sides of each steak, pressing down with your fingertips to ensure the rub binds with the beef.

4.   Place two steaks in each of two food-safe bags, ensuring they are not touching, and seal with your vacuum sealer or the water displacement method. Add the sealed bags to the water bath and cook for 1 hour, 30 minutes.

5.   When the steaks are cooked, carefully remove them, place on a paper towel-lined plate or baking sheet, and pat them dry. Discard the cooking liquid.

6.   Heat a cast iron pan over high heat and add the butter. When the butter begins to bubble, place the steaks in the pan and sear for 1 minute per side.

7.   Let rest for 2 to 3 minutes before slicing into 1-inch strips.

DOUBLE-SEARED CHUCK TAILS

Prep time: 10 MINUTES | Cook time: 20 HOURS | Finishing time: 10 MINUTES | **131°F (55°C)**

I'd never heard of chuck tails before but stumbled across them in Liberty Village in Toronto. Chuck tails are tough cuts, but they have very good marbling through-out. They don't require as long of a cook as the chuck roast, and resemble a New York strip loin in size and shape. I recommend presearing as the cut cooks at a lower temperature for a longer period of time. It also helps to get a thicker bark as the meat doesn't cook as much when it is seared cold versus searing after cooking when it has hit its cooking temperature. Serve with Asiago Broccoli and Peppers (page 58), Garlic and Herb Cheese Cauliflower Purée (page 56) or Spicy Garlic Green Beans (page 47). **4 SERVINGS**

4 tablespoons canola oil, divided
4 chuck beef tail flaps
1 teaspoon garlic powder
1 teaspoon onion powder

1 teaspoon dried basil
1 teaspoon kosher salt
1 teaspoon freshly ground black pepper

1. Set your sous vide circulator to 131°F (55°C).

2. Heat 2 tablespoons of canola oil in a cast iron pan over high heat. Carefully add the tails to the pan and sear for 1 minute per side, rotating 90 degrees after 30 seconds on each side. Remove the tails from the pan and place on a baking sheet to cool.

3. Mix together the garlic powder, onion powder, basil, salt, and pepper in a small bowl. Rub on both sides and ends of the tails.

4. Add the tails to a food-safe bag and seal using your vacuum sealer or the water displacement method. Add the sealed bag to the water bath and cook for 20 hours.

5. When the tails are cooked, remove them from the bag and place on a paper towel–lined plate or baking sheet. Discard the cooking liquids.

6. Heat the remaining 2 tablespoons of oil in a cast iron pan over high heat.

7. When the oil begins to smoke, place the tails in the pan and sear for 30 seconds per side. Let the tails rest for 2 to 3 minutes before plating.

BONE-IN LEG OF LAMB

Prep time: 15 MINUTES | Cook time: 8 HOURS | Finishing time: 15 MINUTES | **133°F (56.1°C)**

Lamb is a popular dish in my household, and every year at Easter, we like to go big. We use a 5-pound leg of lamb, which feeds fourteen people with leftovers—a large undertaking, but the results are huge! If you want to be cautious, start with a much smaller leg to make sure you like it. The cooking time won't change—you can cook a leg of lamb for between 5 and 30 hours—but the tenderness of the lamb will be impacted. The longer you cook, the more it will break down and become softer. Eight hours is my preference. **14 SERVINGS**

10 garlic cloves, minced

2 tablespoons dried rosemary

5 teaspoons extra-virgin olive oil

1 (5-pound) bone-in leg of lamb

1. Set your sous vide circulator to 133°F (56.1°C).

2. Mix together the garlic, rosemary, and olive oil in a bowl. Rub the mix all over the leg of lamb, getting into crevices where possible and along the bone.

3. Place the leg in a large food-safe bag and seal using your vacuum sealer or the water displacement technique. Place the sealed bag in the water bath for 8 hours.

4. When the lamb is cooked, remove it from the bag and reserve the cooking liquid.

5. Heat the grill to between 600°F and 800°F.

6. Place the lamb on the grill and use a baster to baste the leg with the cooking liquid. (Note: The juices may cause flare-ups, so be careful of burning.)

7. "Roll" the leg around the grill to ensure even browning/searing for about 10 minutes.

8. Remove from the grill and let rest for 5 minutes before carving.

MOROCCAN HARISSA LAMB KEBABS

Prep time: 15 MINUTES | Cook time: 8 HOURS | Finishing time: 15 MINUTES | **135°F (57.2°C)**

If you can find Moroccan lamb, you'll notice that it doesn't have the same flavor as New Zealand or US lamb. Moroccan sheep store most of their fat in their tails, so the meat doesn't have as pungent a flavor. These kebabs are packed with flavor, and instead of marinating beforehand like you would with traditional cooking methods, we get the benefit of the lamb marinating while it cooks!

4 TO 6 SERVINGS, 8 TO 12 KEBABS

3 tablespoons extra-virgin
  olive oil, divided
4 teaspoons red wine vinegar
2 tablespoons harissa paste
2 large garlic cloves, minced, plus 1
  teaspoon pressed garlic
1½ teaspoons ground cumin
1½ teaspoons ground coriander
1 teaspoon cayenne pepper

½ teaspoon sea salt
1½ pounds boneless leg or shoulder of
  lamb cut into 1- to 1½-inch cubes
1 cucumber, peeled and finely chopped
Zest of ½ lemon
1 tablespoon freshly squeezed
  lemon juice
1 cup Greek-style yogurt

1. Set your sous vide circulator to 135°F (57.2°C). If using wooden skewers, fill a large bowl of water and soak 12 while the lamb cooks.

2. Mix together 2 tablespoons of olive oil, the vinegar, harissa, minced garlic, cumin, coriander, cayenne, and salt in a bowl.

3. Add the lamb cubes to a food-safe bag in a single layer, or split between two smaller bags if necessary. Pour the marinade into the bag or evenly between the two bags and seal using your vacuum sealer or the water displacement method. Add the sealed bag to the water bath and cook for 2 hours.

4. When the lamb is cooked, remove the cubes and place on a paper towel–lined plate or baking sheet. Discard the cooking liquid.

5. Mix the cucumber, lemon zest and juice, yogurt, and pressed garlic in a small bowl. Set aside.

6. Slide 4 to 5 cubes onto each skewer.

7. Add the remaining 1 tablespoon of oil to a cast iron pan over high heat. Add the skewers, turning often, until they begin to brown, 1 to 2 minutes.

8. Remove and serve with the lemon-garlic yogurt for dipping.

BISON TENDERLOIN

Prep time: 10 MINUTES | Cook time: 3 HOURS | Finishing time: 10 MINUTES | **135°F (57.2°C)**

Not as common as beef tenderloin, bison is typically raised on ranches and grass-fed. Nutritionally, bison has fewer calories than beef but similar levels of protein, vitamins, and iron. Though it is sometimes hard to find, more bespoke butchers are starting to carry it. It tends to be pricier, but it cooks beautifully and predictably sous vide, making the expense worth it. Its flavor profile is leaner and sweeter than beef and will require an extra hour or so to cook perfectly, so plan ahead. Serve with Spicy Garlic Green Beans (page 47), Crispy-Skin Smashed Potatoes (page 49) or Garlic and Herb Cheese Cauliflower Purée (page 56). **4 SERVINGS**

2¼ pounds bison tenderloin

1 teaspoon kosher salt

1 teaspoon freshly ground black pepper

1 tablespoon chili oil

2 teaspoons dried oregano

1 teaspoon garlic powder

½ teaspoon onion powder

½ teaspoon chili powder

2 tablespoons unsalted butter

1. Set your sous vide circulator to 135°F (57.2°C).

2. Season the tenderloin on all sides with the salt and pepper.

3. Mix together the chili oil, oregano, garlic powder, onion powder, and chili powder in a small bowl. Rub the paste all over the tenderloin.

4. Place the tenderloin into a food-safe bag and seal with your vacuum sealer or using the water displacement method. Add the sealed bag to the water bath and cook for 3 hours.

5. When the bison is cooked, transfer it to a paper towel–lined plate or baking sheet.

6. Add the butter to a cast iron pan over high heat. When the butter begins to foam, place the bison in the pan and sear on each side for 45 seconds for a total of 3 minutes.

7. Remove from the heat and let rest for 5 minutes before carving.

SKIRT STEAK *and* EGGS

Prep time: 10 MINUTES | Cook time: 1 HOUR, 30 MINUTES | Finishing time: 15 MINUTES | 131°F (55°C)

Skirt steak, very similar to flatiron or flank steaks, is considered a tough cut. The difference is that skirt is traditionally very thin. That makes it challenging to cook to a rare or medium-rare doneness, but we can remove that worry when cooking sous vide. This is one of the most flavorful beef recipes, absorbing the sugar and umami flavors deep into the beef. Remember, when carving, cut against the grain. This dish goes well with steak sauce and roasted tomatoes. **4 SERVINGS**

1 cup soy sauce
½ cup brown sugar
⅓ cup extra-virgin olive oil
4 garlic cloves, chopped
1 tablespoon sea salt

1 tablespoon freshly ground black pepper
2½ pounds skirt, flatiron, or flank steak
4 eggs
¼ cup milk
½ teaspoon kosher salt

1. Set your sous vide circulator to 131°F (55°C).

2. Mix together the soy sauce, brown sugar, olive oil, garlic, sea salt, and pepper in a small bowl.

3. Place the steak in a large food-safe bag, ensuring it lies flat. Pour in the marinade and seal using your vacuum sealer or the water displacement method. Add the sealed bag to the water bath and cook for 1 hour, 30 minutes.

4. Combine the eggs, milk, and kosher salt and whisk until blended.

5. Heat a small greased skillet over medium-high heat, pour the egg mix into the skillet, and scramble the eggs.

6. When the steak is cooked, remove it and place on a paper towel–lined plate or baking sheet. Discard the cooking liquid.

7. Heat a cast iron pan over high heat and sear the steak for 30 seconds per side.

8. Carve against the grain into 1-inch strips. Serve with the scrambled eggs.

ROSEMARY-GARLIC RACK OF LAMB

Prep time: 5 MINUTES | Cook time: 1 HOUR, 15 MINUTES | Finishing time: 10 MINUTES | 131°F (55°C)

When my mother used to ask me what I wanted for a special dinner, celebrating something important, there was a time when all I would ask for is rack of lamb. Lamb, in general, is an expensive red meat with the most expensive cut being the rib rack. It absorbs aromatic and garlic flavors very well and when served with mint is a mouthwatering combination. This dish is delicious served with mint jelly. **4 SERVINGS**

2 racks of lamb, frenched
½ teaspoon kosher salt
½ teaspoon freshly ground black pepper
1 tablespoon extra-virgin olive oil
1 tablespoon canola oil

2 tablespoons unsalted butter
4 garlic cloves, minced, divided
4 fresh rosemary sprigs, halved, divided
Coarse salt

1. Set your sous vide circulator to 131°F (55°C).

2. Season the racks of lamb on both sides with the salt and pepper.

3. Place the racks in one large, or two medium, food-safe bags and evenly distribute the olive oil all over the lamb. Add the sealed bag to the water bath and cook for 1 hour, 15 minutes.

4. When the lamb is cooked, remove it from the bag, place on a paper towel–lined plate or baking sheet, and pat dry.

5. Heat a cast iron pan over high heat and add the canola oil. When the oil is shimmering, place one rack of lamb meat-side down in the pan and sear for 1 minute. Add 1 tablespoon of butter. When the butter melts, add half the garlic and half the rosemary. Tilt the pan, then spoon butter on top of the lamb. Turn the lamb over and sear for 1 additional minute, repeating the butter basting.

6. Remove the first rack to a cutting board and place the second rack in the pan. Repeat the process of searing and butter basting using the remaining half of the garlic and rosemary.

7. Transfer the second rack to a cutting board and carve both into single servings from where the bones point upward. Serve 3 or 4 pieces per person and season with coarse salt.

BRIE-STUFFED BEEF BURGERS

Prep time: 10 MINUTES | Cook time: 30 MINUTES | Finishing time: 10 MINUTES | 135°F (57.2°C)

I almost always opt for a cheeseburger when I eat out. When I started making (better!) sous vide burgers at home, I realized I had an opportunity to stuff the burgers before cooking. After trying different cheeses, I discovered that Brie worked very well. It bakes well, doesn't liquefy when heat is added, and adds a rich and creamy texture to the meat. **4 SERVINGS**

1 pound lean ground beef
1 yellow onion, finely chopped
1 egg
2 teaspoons Worcestershire sauce
½ teaspoon kosher salt

½ teaspoon freshly ground black pepper
2½ ounces Brie, cut into 4 thin slices
4 fresh white or whole-wheat burger buns
Iceberg lettuce leaves
4 tomato slices

1. Set your sous vide circulator to 135°F (57.2°C).

2. Mix the beef, onion, egg, and Worcestershire sauce with your hands until well blended. Shape the beef mixture into 8 evenly sized patties and season with the salt and pepper.

3. Place 1 Brie slice in the center of each patty and place a second patty over the Brie. Press gently to create a single patty.

4. Add a maximum of two patties to each of four food-safe bags and seal using your vacuum sealer or the water displacement method. Add the sealed bags to the water bath and cook for 30 minutes.

5. When the burgers are cooked, carefully remove them from the bag and place them on a paper towel–lined plate or baking sheet. Discard the cooking liquid.

6. Heat a cast iron pan over high heat. When the pan is hot, add the patties and sear for 1 minute per side.

7. Serve the burgers on a fresh bun with lettuce, tomato, and your choice of condiments.

INDOOR BEEF BRISKET

Prep time: 10 MINUTES | Cook time: 30 HOURS | Finishing time: 3 HOURS, 40 MINUTES | 155°F (68.3°C)

This was my first sous vide video post on social media. It garnered quite a bit of attention (and some scrutiny from the meat smokers out there) because it had all the components of a smoked brisket: a nice dark bark, juicy texture, a smoke ring (using pink curing salt), and a smoky flavor. It tasted like a smoked brisket— even though we were cheating a little bit by using liquid smoke—and it was made entirely indoors. This will work well for those of you who don't have access to a smoker, or even outdoor space. It's one of the longest cooks in this section of the book, but the results are worth it! **8 TO 10 SERVINGS**

8 tablespoons brown sugar
2 tablespoons kosher salt
2 teaspoons freshly ground black pepper
2 teaspoons paprika
1 teaspoon cayenne pepper

1 teaspoon dried parsley
1 teaspoon garlic powder
½ teaspoon ground cumin
4 pounds first-cut brisket
¼ teaspoon liquid hickory smoke

1. Set your sous vide circulator to 155°F (68.3°C).

2. Mix together the brown sugar, salt, pepper, paprika, cayenne, parsley, garlic powder, and cumin in a medium bowl. Reserve one-quarter of the mixture, and rub the rest of the mixture evenly all over the brisket.

3. Place the brisket in a large food-safe bag, add the liquid smoke, and seal using your vacuum sealer or water displacement method. Add the sealed bag to the water bath and cook for 30 hours. Ensure you place a lid or other covering over the cooking container, as water will evaporate significantly during this long cook.

4. When the brisket is cooked, remove it from the water bath and let it sit for at least 1 hour in the bag.

5. Preheat the oven to 300°F.

6. Remove the brisket from the bag and place it on a large baking sheet. Discard the cooking liquid. Pat the brisket dry with paper towels, and rub the reserved seasoning mixture all over the brisket.

7. Place the brisket on a wire rack over a large baking sheet and roast until a brown bark has formed, about 2 hours.

8. Remove the brisket from the oven and cover with aluminum foil for 40 minutes.

9. Carve against the grain and serve with baked beans, fresh bread, and butter.

VEAL MARSALA

Prep time: 10 MINUTES | Cook time: 1 HOUR, 45 MINUTES | Finishing time: 10 MINUTES | 140°F (60°C)

Veal marsala is an Italian American dish that leverages the flavors of Marsala wine. It doubles as a chicken dish (following the chicken temperature and time guide on page 200) but really performs well when cooked sous vide. The tenderness of veal is amplified when cooked sous vide and the rich flavors make it a delicate beef dish that pairs well with both red and white wines. **4 TO 6 SERVINGS**

¾ cup chicken broth
½ cup sweet Marsala
¼ cup sliced portobello mushrooms
3 tablespoons unsalted butter
3 tablespoons extra-virgin olive oil
4 garlic cloves, pressed

1 shallot, finely chopped
½ teaspoon kosher salt
½ teaspoon freshly ground black pepper
8 veal cutlets
1 fresh rosemary sprig
4 cups steamed white rice

1. Set your sous vide circulator to 140°F (60°C).

2. Combine the broth, Marsala, mushrooms, butter, olive oil, garlic, shallot, salt, and pepper in a medium bowl.

3. Place the veal cutlets in a single layer in a large food-safe bag, or use two smaller bags if necessary. Place the rosemary over the cutlets and add the marinade, dividing it evenly if using two bags. Add the sealed bag to the water bath and cook for 1 hour, 45 minutes.

4. When the veal is cooked, remove the cutlets from the bag and place on a paper towel–lined plate or baking sheet.

5. Discard the rosemary and transfer the cooking liquid to a large saucepan. Bring to a low boil over medium heat and simmer for 5 minutes.

6. Add the cutlets to the pan after the sauce has thickened, and simmer for 1 minute.

7. Remove the cutlets and place over steamed rice, pour the reduced sauce over the veal and rice, and serve.

ROSEMARY VEAL CHOPS

Prep time: 10 MINUTES | Cook time: 3 HOURS | Finishing time: 10 MINUTES | 134°F (56.6°C)

An impressive and sometimes very large cut, the veal chop looks similar to a tomahawk or barbarian cut rib steak. The texture, however, is different. Veal is much tenderer due to its shorter life and lack of developed muscle and connective tissue. It absorbs flavors easily and responds well to sous vide cooking methods. Cook with the bone in for an impressive table presence, but make sure you don't poke a hole in the cooking bag. Serve with Spicy Garlic Green Beans (page 47) or Sweet Potato Fries with Malt Vinegar Mayonnaise (page 50). **6 SERVINGS**

6 tablespoons unsalted butter

4 tablespoons dry white wine

4 tablespoons low-sodium chicken broth

4 fresh rosemary sprigs

2 garlic cloves, pressed

1½ teaspoons finely chopped fresh thyme

6 (¾-pound or 1-pound if bone-in) veal chops

1 teaspoon kosher salt

1 teaspoon freshly ground black pepper

2 tablespoons extra-virgin olive oil

1. Set your sous vide circulator to 134°F (56.6°C).

2. Combine the butter, white wine, broth, rosemary, garlic, and thyme in a medium saucepan over medium-high heat and cook for 5 minutes. Remove from the heat and let cool.

3. Place the veal chops on a clean flat surface and season with the salt and pepper.

4. Add a maximum of two chops to each of three food-safe bags and divide the butter-rosemary mixture evenly between the bags. Add the sealed bags to the water bath and cook for 3 hours.

5. When the chops are cooked, remove them from the bags and place on a paper towel–lined plate or baking sheet.

6. Heat a cast iron pan over high heat. Brush the chops on all sides with the olive oil. Add the chops to the pan and sear them for 45 seconds per side.

7. Transfer the chops to a cutting board and let rest for 5 minutes.

TEVIN SPICY GOAT CURRY

Prep time: 10 MINUTES | Cook time: 30 HOURS | Finishing time: 20 MINUTES | 180°F (82.2°C)

When on vacation in Jamaica last summer, I was lucky enough to have some authentic Jamaican cuisine. Also popular in India, goat curry is highly flavorful and often very spicy. This dish incorporates the hot Scotch bonnet pepper, but you can substitute a milder pepper if you want to scale things back a little bit. Though it is a long cook, the flavors make the wait very much worth it. **4 SERVINGS**

2 tablespoons ghee, divided
2 Scotch bonnet peppers, chopped
3 garlic cloves, minced, divided
1 teaspoon turmeric
1 teaspoon ground cumin
1 teaspoon smoked paprika
1 teaspoon grated fresh ginger
½ teaspoon kosher salt
2 cardamom pods

2 fresh thyme sprigs
2¼ pounds boneless goat meat, cut into large pieces
1 large onion, chopped
3 tomatoes, chopped
1 teaspoon garam masala
2 tablespoons full-fat plain yogurt
1 tablespoon chopped fresh cilantro

1. Set your sous vide circulator to 180°F (82.2°C).

2. Mix together 1 tablespoon of ghee, peppers, 2 minced garlic cloves, turmeric, cumin, paprika, ginger, salt, cardamom, and thyme in a bowl.

3. Place the goat in one large food-safe bag, or use two medium bags if necessary, pour the ghee mixture evenly over the goat, and seal using your vacuum sealer or the water displacement method. Add the sealed bag to the water bath and cook for 30 hours.

4. When the goat is cooked, remove it from the water bath and set aside.

5. Heat the remaining 1 tablespoon of ghee in a large saucepan over high heat. Add the onion and cook for 4 minutes, stirring often. Add the remaining 1 clove of minced garlic and cook for 1 additional minute.

6. Add the tomatoes and garam masala and reduce the heat to medium. Simmer for 2 minutes, then add the yogurt followed by the goat and cooking liquid. Simmer for 10 to 15 minutes.

7. Serve over a bed of rice and garnish with the chopped cilantro.

HAVANA BEACHFRONT TACOS

Prep time: 5 MINUTES, PLUS 30 MINUTES MARINATING TIME | Cook time: 90 MINUTES
Finishing time: 5 MINUTES | 131°F (55°C)

Havana is on my list of places to visit, but until then, I'll find ways to enjoy it from abroad. This recipe also gives me an opportunity to cook flank steak, one of my favorite cuts of beef to cook sous vide, and combine some citrus and spice flavors to bring Havana home. **4 SERVINGS, 3 TACOS PER PERSON**

2 pounds flank steak
2 teaspoons kosher salt
2 teaspoons freshly ground black pepper
1 teaspoon garlic powder
2 teaspoons freshly squeezed lime juice
Zest of 1 lime
2 teaspoons freshly squeezed
   orange juice
Zest of 1 orange

1 teaspoon red pepper flakes
1 garlic clove, pressed
1 tablespoon canola oil
1 tablespoon unsalted butter
12 corn tortillas
1 head cabbage, sliced, for serving
Pico de gallo, for serving
Sour cream, for serving
4 jalapeño peppers, sliced, for serving

1. Set your sous vide circulator to 131°F (55°C).

2. Lay the flank steak on a clean flat surface and season with salt, pepper and garlic powder on both sides.

3. Combine the lime juice and zest, orange juice and zest, red pepper flakes, and garlic in a bowl and mix thoroughly.

4. Add the steak and marinade to a food-safe bag and seal using your vacuum sealer or using the water displacement method. Refrigerate for at least 30 minutes and up to 24 hours. Place the sealed bag in the water bath and cook for 90 minutes.

5. When the steak is cooked, remove it from bag, place on a paper towel–lined plate or baking sheet, and pat it dry. Discard the cooking liquid.

6. Heat the oil in a cast iron pan over high heat and add the oil and butter. Sear the steak for a total of 1 minute, flipping every 15 seconds.

7. Remove from the heat and slice against the grain. Place a few slices of steak in each tortilla and garnish with cabbage, pico de gallo, sour cream, and jalapeños and serve immediately.

CURRYWURST BEEF SAUSAGES

Prep time: 5 MINUTES | Cook time: 90 MINUTES | Finishing time: 10 MINUTES | **135°F (57.2°C)**

This recipe is similar to the No-Grill Summer Sausages (page 101), but with two changes: 1) the recipe uses beef sausages, and 2) I recommend you eat them currywurst-style—sliced up into bite-size pieces and served with curry ketchup or hot mustard for dipping. I frequent a restaurant called WVRST, in Toronto, which has perfected this style of sausage eating. Now you can enjoy it at home.

4 SAUSAGES

4 beef link sausages
½ cup hot mustard, for dipping
½ cup curry ketchup, for dipping

1. Set your sous vide circulator to 135°F (57.2°C).

2. Add the sausages to a food-safe bag and seal using your vacuum sealer or the water displacement method. Add the sealed bag to the water bath and cook for 90 minutes.

3. Heat a cast iron pan over high heat, or heat your grill to high (400°F to 600°F). Place the sausages in the pan or on the grill and cook until cooked through and showing grill marks (if cooked on grill), 1 to 3 minutes.

4. Slice and serve with the hot mustard and curry ketchup for dipping.

Tip: Because the sausages are very soft when raw, using the water displacement method rather than a vacuum sealer is recommended to help preserve the original shape of the sausage and avoid warping.

ITALIAN MEATBALLS

Prep time: 20 MINUTES | Cook time: 90 MINUTES | Finishing time: 5 MINUTES | 140°F (60°C)

My apologies to any Italian grandmothers reading this book—I know that meatballs cooked in a bag can't beat your homemade recipe. That being said, if you don't have an Italian grandmother to make you meatballs, you're in luck! This recipe creates perfect meatballs, enjoyed on their own or incorporated into a pasta dish, and they are fun to make with your little ones. **4 SERVINGS, 14 TO 16 MEATBALLS**

1 pound lean ground beef
½ cup white bread crumbs
¼ cup milk
1 egg, beaten
2 tablespoons finely chopped
   fresh parsley

1 garlic clove, pressed
1 teaspoon salt
½ teaspoon dried basil
1 tablespoon canola oil

1. Set your sous vide circulator to 140°F (60°C).

2. Combine the beef, bread crumbs, milk, egg, parsley, garlic, salt, and basil in a medium bowl and mix with your hands until thoroughly blended.

3. Working in batches, use your hands to roll 14 to 16 meatballs, placing 4 to 6 balls in a single layer into each food-safe bag.

4. Seal using your vacuum sealer on medium or using the water displacement method. Be careful not to squeeze the meatballs too much during the vacuum process to maintain their round shape. Place the sealed bags in the water bath and cook for 90 minutes.

5. Heat the oil in a large cast iron pan over medium-high heat.

6. When the meatballs are cooked, use a slotted spoon to carefully remove them from the bags. Discard the cooking liquid. Sear the meatballs in the hot pan until browned, 4 to 5 minutes.

7. Remove from the heat and serve immediately or store in the refrigerator for 1 to 2 days.

SOUTH AFRICAN SOSATIES

Prep time: 20 MINUTES | Cook time: 8 HOURS | Finishing time: 20 MINUTES | 140°F (60°C)

My parents were born in South Africa, and as a family, we traveled there yearly until I turned 16. Since then, I've returned only once. Growing up, I only knew kebabs by their South African name—sosaties. It had been so long since I'd had kebabs like those I'd eaten as a kid, that when I started experimenting more with sous vide, sosaties popped into my mind. I already knew that lamb cooks well under vacuum, so it was simply a matter of getting input from my family on a perfect recipe. Serve the skewers with basmati rice or Garlic-Parmesan Asparagus (page 57) or Mint Green Pea Purée (page 48). Smaaklike ete! (Bon appétit!) **6 SERVINGS**

¾ cup white wine vinegar
½ cup dry red wine
2 medium yellow onions, chopped
4 garlic cloves, pressed
Zest of 2 lemons
6 tablespoons brown sugar
2 tablespoons coriander seeds, crushed
1 tablespoon apricot jam
1 tablespoon corn flour
1 tablespoon curry powder

1 tablespoon grated fresh ginger
2 teaspoons kosher salt
1 teaspoon ground allspice
1 teaspoon ground cinnamon
4½ pounds lamb shoulder, cut into
  1-inch cubes
1 tablespoon unsalted butter
6 pearl onions, peeled and halved
12 dried apricots, halved
2 tablespoons extra-virgin olive oil

1. Set your sous vide circulator to 140°F (60°C). If using wooden skewers, fill a large bowl with water and soak 12 to 16 skewers while the lamb is cooking.

2. Combine the vinegar, red wine, yellow onions, garlic, lemon zest, brown sugar, coriander, apricot jam, corn flour, curry powder, ginger, salt, allspice, and cinnamon in a large bowl and mix.

3. Place the lamb in a large food-safe bag in a single layer, or divide between two smaller bags if necessary. Cover with the marinade and seal using your vacuum sealer or the water displacement method. Add the sealed bag to the water bath and cook for 8 hours.

4. About 20 minutes before the lamb is cooked, melt the butter in a small saucepan. Sauté the pearl onions until softened and brown, 8 to 10 minutes. Remove from the heat and place on a paper towel–lined plate to cool.

5. Remove the lamb from the cooking bag and place on a paper towel–lined plate or baking sheet. Pat it dry.

6. Pour the cooking juices into a medium saucepan over medium heat and cook until reduced by half, 10 minutes.

7. Slide the kebab ingredients onto each skewer, alternating the ingredients: lamb, onion, lamb, apricot, lamb, apricot, onion, lamb, or whatever sequence you prefer.

8. Heat the oil in a cast iron pan over high heat, or heat a grill to high. Cook or grill the kebabs 45 seconds on each side.

9. Drizzle the reduced marinade over the kebabs, and serve

BEEF BONE MARROW

Prep time: 5 MINUTES | Cook time: 1 HOUR | Finishing time: 5 MINUTES | **155°F (68.3°C)**

I only recently had the chance to try bone marrow, and it's one of those things I wish I had discovered earlier in life. A beef leg bone, cut in half with a bone saw and then broiled until the marrow liquefies, may not sound appetizing, but I assure you it is. Cooking it sous vide helps manage the transformation from hard to soft marrow, and it only requires a few minutes in the oven to roast at the very end. **8 SERVINGS**

4 large beef marrowbones, canoed (making 8 halves)

½ teaspoon kosher salt

¼ cup chopped fresh parsley

1. Set your sous vide circulator to 155°F (68.3°C).

2. Sprinkle the bone marrow canoes evenly with the salt.

3. Place the marrowbones into two large or four small food-safe bags and seal using your vacuum sealer or the water displacement method. Add the sealed bags to the water bath and cook for 1 hour.

4. Remove the bones from the cooking bags and place bone-side down on a large baking sheet.

5. Set the oven to broil and place the baking sheet on the middle rack of the oven. Broil until browned, 2 to 4 minutes.

6. Remove from the oven and sprinkle with the parsley.

7. Serve with toasted bread or scoop marrow directly onto beef dishes.

SESAME-HOISIN BRAISED BEEF SHORT RIBS

Prep time: 5 MINUTES | Cook time: 12 HOURS | Finishing time: 10 MINUTES | 185°F (85°C)

Selecting the menu for our wedding, my wife and I were lucky enough to have two tastings to make sure we were satisfied. Both times, chef Stephen Treadwell wowed us with his sesame-hoisin braised beef ribs. I knew I wanted to make this amazing dish myself. After a few tries I now have it pretty close. This is on the longer end of cooking times so it's best to start cooking early in the morning. These are wonderful served on a bed of Garlic and Herb Cheese Cauliflower Purée (page 56) or traditional mashed potatoes. **6 SERVINGS**

1½ pounds beef short ribs
½ teaspoon kosher salt
½ teaspoon freshly ground black pepper
3 tablespoons toasted sesame oil
1½ cups hoisin sauce
10 to 12 garlic cloves, smashed

3 tablespoons rice wine vinegar
2 tablespoons pressed fresh ginger
2 tablespoons unsalted butter
⅛ cup finely chopped scallions
⅛ cup sesame seeds

1. Set your sous vide circulator to 185°F (85°C).

2. Season the short ribs all over with the salt and pepper.

3. Heat a cast iron pan over high heat, add the sesame oil, and sear the ribs on all sides for a total of 2 minutes.

4. Mix together the hoisin sauce, garlic, vinegar, and ginger in a small bowl.

5. Place a maximum of three ribs in a food-safe bag, cover with the hoisin sauce mixture, and seal using your vacuum sealer or the water displacement method. Add the sealed bag to the water bath and cook for 12 hours.

6. When the beef is cooked, remove the ribs from the cooking bag and place them on a paper towel–lined plate or baking sheet.

7. Pour the cooking liquid into a small saucepan over medium-high heat for 4 to 5 minutes until sticky.

8. Heat a cast iron pan over high heat, and melt the butter until foaming. Add the short ribs and sear for 1 minute per side.

9. Transfer the ribs to a plate and spoon the hoisin sauce over the top. Garnish with the scallions and sesame seeds.

KOREAN KALBI BEEF RIBS

Prep time: 10 MINUTES, PLUS 2 HOURS MARINATING TIME | Cook time: 3 HOURS
Finishing time: 10 MINUTES | 140°F (60°C)

These short ribs are exactly that—short. They are crosscut beef ribs and are very thin so you can enjoy the flavor of beef ribs without the 12-hour wait. The dish is a Korean dish featuring brown sugar, sesame, rice wine, and orange, and the meat pulls right off the bone, leaving your plate spotless apart from the oval remains of the disk. **4 TO 6 SERVINGS**

12 to 16 cross-cut beef ribs
  (about 3 pounds)
¼ teaspoon kosher salt
¼ teaspoon freshly ground black pepper
½ cup brown sugar
½ cup soy sauce
¼ cup mirin (rice wine)

¼ cup freshly squeezed orange juice
2 tablespoons sesame oil
2 tablespoons minced garlic
1 teaspoon red pepper flakes
¼ cup chopped scallions
¼ cup sesame seeds

1.   Set your sous vide circulator to 140°F (60°C).

2.   Place the ribs on a clean surface and season them all over with the salt and pepper.

3.   Whisk together the brown sugar, soy sauce, mirin, orange juice, sesame oil, garlic, and red pepper flakes in a small bowl.

4.   Place the ribs flat, in a single layer, in two large food-safe bags. Pour the marinade evenly across the bags and seal using your vacuum sealer or the water displacement method. Refrigerate the ribs for 2 hours to marinate.

5.   Place the sealed bags in the water bath and cook for 3 hours.

6.   When the ribs have finished cooking, remove them from the bags, place them on a paper towel–lined baking sheet, and pat them dry. Remove the paper towel from the sheet.

7.   Heat the cooking liquid in a small saucepan over high heat until thickened, 4 to 5 minutes.

8.   Set the oven to broil. Use a soft brush to brush the reduced marinade all over the beef ribs. Broil for 1 to 2 minutes per side.

9.   Remove from the oven and sprinkle with the scallions and sesame seeds before serving.

LOMO SALTADO

Prep time: 10 MINUTES | Cook time: 1 HOUR, 30 MINUTES | Finishing time: 10 MINUTES | **133°F (56.1°C)**

While on my honeymoon in Peru, I took two cooking classes and learned about this dish, which uses alpaca. I was a bit wary, but it turns out alpaca tenderloin is incredibly tender and flavorful. Tracking down alpaca outside of Peru might be tricky, so beef sirloin makes a great replacement for this take on the traditional dish. **4 TO 6 SERVINGS**

3 pounds beef sirloin
½ tablespoon kosher salt
½ tablespoon freshly ground
   black pepper
2 tablespoons extra-virgin olive oil,
   divided
2 tablespoons white wine vinegar

½ tablespoon freshly squeezed lime juice
½ tablespoon garlic powder
1 large onion, finely chopped
1 large tomato, finely chopped
2 garlic cloves, pressed
2 tablespoons reduced-sodium soy sauce
4 cups steamed white (or any other) rice

1.   Set your sous vide circulator to 133°F (56.1°C).

2.   Use a sharp knife to cut the sirloin into 1-inch strips. Season the slices with the salt and pepper.

3.   Mix together 1 tablespoon of olive oil, white wine vinegar, lime juice, and garlic powder in a bowl.

4.   Add the sirloin to the marinade and toss to coat, then transfer the meat and marinade to a large food-safe bag in a single layer and seal using your vacuum sealer or the water displacement method. Add the sealed bag to the water bath and cook for 1 hour, 30 minutes.

5.   Heat a large saucepan over medium-high heat and add the remaining 1 tablespoon of olive oil. Add the onion, tomato, garlic, and soy sauce and stir until the tomatoes begin to soften, about 5 minutes. Remove from the heat.

6.   When the alpaca has finished cooking, remove the strips, place them on a paper towel–lined plate or baking sheet, and pat them dry. Reserve the cooking liquids.

7.   Heat a cast iron pan over high heat and add the beef strips to the pan. Sear quickly, 1 to 1½ minutes.

8.   Add the cooking liquids to the tomato-onion mix and heat on high until simmering, 4 to 5 minutes. Add the beef strips to the sauce and toss for an additional 2 minutes. Serve on a bed of your favorite rice.

OREGON RUBBED TRI-TIP

Prep time: 10 MINUTES | Cook time: 3 HOURS | Finishing time: 10 MINUTES | **133°F (56.1°C)**

Tri-tip beef isn't the most commonly found steak on restaurant menus, in the local grocery store, or even at your butcher's counter. It is on the tougher end of the tender meat spectrum but it locks in flavors well. It typically comes in the shape of a small roast, so it will feed a group of four to six very well. The rub here incorporates a variety of flavors including dried aromatics, onion, and paprika. Serve with Crispy-Skin Smashed Potatoes (page 49) or Sous Vide Buttered Corn on the Cob (page 55). **4 TO 6 SERVINGS**

3 pounds beef tri-tip
2 tablespoons extra-virgin oil
1½ teaspoons garlic powder
½ tablespoon salt
¼ teaspoon ground black pepper
¼ teaspoon onion powder

¼ teaspoon paprika
¼ teaspoon dried dill
¼ teaspoon dried sage
¼ teaspoon crushed dried rosemary
2 tablespoons unsalted butter

1. Set your sous vide circulator to 133°F (56.1°C).

2. Using a soft brush, cover the tri-tip with the olive oil.

3. Mix together the garlic powder, salt, pepper, onion powder, paprika, dill, sage, and rosemary in a bowl. Rub the spice mixture all over the beef, ensuring an even coating.

4. Place the tri-tip in a large food-safe bag and seal using your vacuum sealer or the water displacement method. Add the sealed bag to the water bath and cook for 3 hours.

5. When the beef is cooked, remove it from the bag and place on a paper towel–lined plate or baking sheet.

6. Heat a cast iron pan over high heat and add the butter. When the butter starts to foam, gently place the tri-tip in the pan and sear on all sides, 2 to 3 minutes total.

7. Let it rest for 5 minutes, then slice against the grain into thin slices.

VENISON BACKSTRAP

Prep time: 5 MINUTES | Cook time: 2 HOURS, 30 MINUTES | Finishing time: 10 MINUTES | 131°F (55°C)

For the hunters out there, this one is for you. Typically, venison has a gamy flavor, so it isn't as popular as more traditional cuts of red meat. It does, however, work very well when cooked sous vide. It is tender, packed full of flavor, and not overpoweringly gamy. Serve with Crispy-Skin Smashed Potatoes (page 49) or Sous Vide Buttered Corn on the Cob (page 55). **4 TO 6 SERVINGS**

3 pounds venison backstrap or loin
2 tablespoons extra-virgin olive oil
2 tablespoons kosher salt
2 tablespoons freshly ground
  black pepper
2 tablespoons brown sugar
2 tablespoons powdered beef bouillon
2 tablespoons onion powder

2 tablespoons garlic powder
1 tablespoon smoked paprika
2 teaspoons cayenne pepper
1 teaspoon dried thyme
1 teaspoon ground cumin
½ teaspoon dried oregano
2 tablespoons butter, unsalted

1. Set your sous vide circulator to 131°F (55°C).

2. Using a soft brush, cover the venison with the olive oil.

3. Mix together the salt, pepper, brown sugar, powdered beef bouillon, onion powder, garlic powder, smoked paprika, cayenne, thyme, cumin, and oregano in a bowl. Rub the spice mixture all over the venison, ensuring an even coating.

4. Place the venison in a large food-safe bag and seal using your vacuum sealer or the water displacement method. Add the sealed bag to the water bath and cook for 2 hours, 30 minutes.

5. When the backstrap is cooked, remove it from the bag and place on a paper towel–lined plate or baking sheet.

6. Heat a cast iron pan over high heat and add the butter. When the butter starts to foam, gently place the venison in the pan and sear on all sides, 2 to 3 minutes total.

7. Let it rest for 5 minutes, then carve against the grain into thin slices.

chapter seven
Drinks and Desserts

Dessert is my favorite course, and I'll eat anything I can that features salted caramel. One of the things about sous vide that surprises people the most is when you serve a sous vide dessert. You can almost see them trying to work it out in their head: "You made cheesecake in your water bath?" Yes, and it's amazing!

Last summer I hosted a sous vide infusion tasting party where we took a variety of standard base alcohols for cocktails—rum, vodka, gin, bourbon, to name a few—and began mixing different flavors to see what worked and what didn't. That tasting party formed the foundation for the cocktails in this chapter, and I'm still experimenting today. Cheers!

LEMON AND BLUEBERRY CRÈME BRÛLÉE (page 165)

LEMON *and* BLUEBERRY CRÈME BRÛLÉE

Prep time: 15 MINUTES, PLUS 5 HOURS COOLING TIME | Cook time: 45 MINUTES
Finishing time: 5 MINUTES | **195°F (90.5°C)**

Crème brûlée is a crowd-pleasing classic French dessert. It can easily be prepared ahead of time and stored in the refrigerator then quickly finished at the dinner table, making it a great option for a dinner party dessert. I love the tartness of the lemon and the freshness of the blueberries as a contrast to the sweet, creamy richness of the crème brûlée, but you can change out the flavors as you prefer—lime, coffee, strawberry, coconut, chocolate—the possibilities are endless! **6 SERVINGS**

6 large egg yolks

1⅓ cups superfine sugar, plus more
   for sprinkling

3 cups heavy (whipping) cream

Zest of 2 lemons

4 tablespoons freshly squeezed
   lemon juice

1 teaspoon vanilla extract

1 cup fresh blueberries

1. Set your sous vide circulator to 195°F (90.5°C).

2. Using an electric mixer, whisk together the egg yolks and sugar in a large bowl until pale and creamy. Set aside.

3. Pour the cream into a medium saucepan over medium heat and heat to just below boiling point. Be careful not to burn the cream. Add the lemon zest, lemon juice, and vanilla to the cream, stirring as you add the lemon juice to avoid curdling. Simmer over low heat for 4 to 5 minutes.

4. Remove the cream mixture from the heat and cool for 2 to 3 minutes. When cooled, pour a small amount into the egg mixture and whisk to combine. (This tempers the egg mixture so it doesn't scramble when you pour in the rest of the hot cream mixture.) Pour the remainder of the cream mixture into the eggs and stir to combine.

5. Divide the blueberries evenly among six mini mason jars, then pour the egg-cream mixture over the blueberries, dividing evenly among the jars.

continued

6. Screw on the jar lids to fingertip tightness (see tip on page 43) and place in the water bath, ensuring they are completely submerged. Cook for 45 minutes.

7. Remove the jars from the water bath and place in the refrigerator for at least 5 hours or up to 4 days.

8. To serve, remove the lids and sprinkle a thin layer of sugar over the top of the crème brûlée. Caramelize the sugar using a blowtorch.

SUMMER BERRY BREAD PUDDING

Prep time: 30 MINUTES | Cook time: 2 HOURS | Finishing time: 2 MINUTES | **170°F (77°C)**

This is a fresh twist on an old English dessert. I have replaced the usual golden raisins, candied peel, and heavier spices with fresh summer berries. This dessert can be eaten hot or cold and would be a great way to finish off a picnic lunch—the mason jars make the pudding easily transportable in a cooler. Alternatively, you could serve this at home hot out of the sous vide topped with Mixed Berry Coulis (page 192) and whipped cream. **6 SERVINGS**

4 slices brioche loaf, cubed, or white
   bread, crusts removed
6 egg yolks
1⅛ cups superfine sugar
2 cups heavy (whipping) cream
1 cup milk

2 teaspoons vanilla extract
1 vanilla pod, halved, seeds scraped
   and reserved
1 cup mixed fresh berries (strawberries,
   raspberries, blackberries) cut into
   small pieces

1. Set your sous vide circulator to 170°F (77°C) and preheat the oven to 350°F (176°C).

2. Spread the bread cubes on a baking sheet and toast in the oven for 5 minutes, watching and turning the cubes as necessary to ensure even browning. (Toasting the bread cubes dries them out, which will allow them to soak up more of the egg-cream mixture later.)

3. Using an electric mixer, whisk together the egg yolks and sugar in a large bowl until pale and creamy.

4. Pour the cream and milk into a saucepan over medium heat and heat to just below boiling point. Be careful not to burn the cream. Add the vanilla, vanilla pod seeds, and vanilla pod to the cream-milk mixture. Simmer over low heat for 4 to 5 minutes, then remove from the heat and cool for 2 to 3 minutes.

5. When cooled, pour a small amount of the cream through a fine mesh sieve into the egg mixture and whisk to combine. (This tempers the egg mixture so it doesn't scramble when you pour in the rest of the hot cream mixture.) Pour the remainder of the cream through the fine mesh sieve into the eggs and stir to combine.

continued

6. Add the bread cubes to the egg-cream mixture and allow the bread to absorb the liquid. After a few minutes, pour any excess liquid that has not been absorbed into a bowl. Set aside.

7. Add the chopped berries to the bread mixture and toss lightly to combine, then divide the bread and berry mixture between six mini mason jars. If there is room in the jars, use the reserved egg-cream mixture to top up the jars.

8. Screw on the jar lids to fingertip tightness and place in the water bath, ensuring they are completely submerged. Cook for 2 hours.

9. If serving immediately, remove the lids and serve. Otherwise, place the jars in the refrigerator and store for up to 3 days.

MOCHA MASON JAR BROWNIES

Prep time: 15 MINUTES | Cook time: 3 HOURS | Finishing time: 2 MINUTES | **195°F (90.5°C)**

This dessert is a decadent chocolate treat. The texture is gooier than a traditional oven-baked brownie, making it something between a brownie and a molten chocolate lava cake. The addition of the coffee products deepens and enhances the chocolate flavors, but you can always omit them if you aren't a coffee fan. For something slightly lighter, you could substitute the fresh strawberries for the dulce de leche sauce. **8 TO 10 SERVINGS**

⅔ cup semisweet chocolate, chopped
8 tablespoons unsalted butter
⅔ cup superfine sugar
2 egg yolks plus 1 egg
2 tablespoons instant coffee powder

1 tablespoon vanilla extract
1 tablespoon coffee liqueur (optional)
½ cup all-purpose flour
Easy Dulce de Leche (page 189) and
    vanilla ice cream, for serving

1. Set your sous vide circulator to 195°F (90.5°C).

2. Melt the chocolate and butter together using a double boiler or a microwave.

3. Whisk the sugar into the chocolate-butter mixture until dissolved. Add the egg yolks, one at a time, and whisk until combined. Add the whole egg and continue whisking until combined. Add the coffee powder, vanilla, and coffee liqueur (if using) and whisk to combine. Add the flour and stir lightly to combine.

4. Pour the chocolate mixture into eight to ten mini mason jars, leaving ¾ inch of space between the top of the jar and the mixture.

5. Screw on the jar lids to fingertip tightness and place in the water bath, ensuring they are completely submerged. Cook for 3 hours.

6. Remove the jars from the water bath and allow to cool for 1 minute. Remove the lids and serve immediately in the mason jars, topped with a spoonful of Easy Dulce de Leche and vanilla ice cream.

SALTED CARAMEL CHEESECAKE

Prep time: 15 MINUTES, PLUS 4 HOURS COOLING TIME | Cook time: 1 HOUR, 30 MINUTES
Finishing time: 10 MINUTES | **176°F (80°C)**

I love salted caramel—salted caramel candy, salted caramel ice cream, even salted caramel vodka! So it's no surprise that salted caramel cheesecake is one of my favorite desserts. You can play around with the ratio of salt to sweet to suit your own taste, although for my palate, the quantities below provide a nice balance. **6 SERVINGS**

2 cups cream cheese,
  at room temperature
3 eggs
1 teaspoon vanilla extract
1 cup Easy Dulce de Leche
  (page 189), divided

⅓ cup heavy (whipping) cream
1 cup graham cracker crumbs
3 tablespoons unsalted butter, melted
½ teaspoon good quality coarse sea salt,
  plus extra for serving

1. Set your sous vide circulator to 176°F (80°C).

2. Using an electric mixer on high speed, whisk the cream cheese, eggs, and vanilla in a bowl until smooth. Add ¾ cup of Easy Dulce de Leche and mix until fully combined. Add the cream and mix until fully combined. Set aside.

3. Combine the graham cracker crumbs and melted butter in another bowl and mix until clumps form.

4. Evenly divide the graham cracker crumbs between six mini mason jars and press firmly with a teaspoon to form a crust in the bottom of each jar.

5. Spoon the cream cheese mixture into the mason jars, dividing evenly.

6. Screw on the jar lids to fingertip tightness and place into the water bath, ensuring they are completely submerged. Cook for 1 hour, 30 minutes.

7. Remove the jars from the water bath and refrigerate for at least 4 hours and up to 3 days.

8. Before serving, stir the salt into the remaining ¼ cup of Easy Dulce de Leche. Remove the jars from the refrigerator, open, and top with a spoonful of salted caramel mixture. Sprinkle a little extra salt on top and serve.

MALT ICE CREAM BASE

Prep time: 15 MINUTES, PLUS TIME REQUIRED TO CHURN ICE CREAM
Cook time: 1 HOUR, 10 MINUTES | Finishing time: 10 MINUTES | 185°F (85°C)

My wife's love of malted milk shakes inspired this recipe, which makes a fantastic addition to a milk shake or a flavorful take on a banana split when you add a few scoops on top of a banana and finish with whipped cream, caramel sauce, and some chopped macadamias and chocolate malt balls. While I have flavored this ice cream base with malt, you can replace it with your own favorite flavors. This recipe gives you the base for the ice cream, and you will still need to churn it using your preferred method for making ice cream. **6 SERVINGS**

1 cup milk
1 cup heavy (whipping) cream
1 teaspoon vanilla extract
6 egg yolks

½ cup Sous Vide Simple Syrup
  (page 190)
½ cup malt powder
Pinch salt

1. Set your sous vide circulator to 185°F (85°C).

2. Using an electric mixer on high speed, whisk together the milk, cream, vanilla, egg yolks, and simple syrup in a bowl until smooth and frothy, 1 to 2 minutes.

3. Place the malt powder into a separate bowl and pour in a small amount of the milk mixture. Stir until the malt becomes liquid, adding a bit more milk mixture as required, and smoothing out all lumps.

4. Pour the malt mixture through a fine mesh strainer into the milk mixture and stir to combine then pour the malt milk mixture into a resealable bag and seal using the water displacement method.

5. Place the bag in the water bath and cook for 1 hour, 10 minutes. Every 10 minutes, use tongs to remove the bag from the water and place on a dish towel on a hard surface. Use oven gloves to massage the bag lightly for 15 seconds and then place the bag back in the water bath. You need to do this to avoid lumps forming in the ice cream base, which would prevent a smooth texture in the final product.

6. Remove the bag from the water bath and immediately place in an ice bath until completely cooled.

7. Follow your preferred method for churning ice cream.

HONEY SPICED PLUMS

Prep time: 20 MINUTES | Cook time: 45 MINUTES | Finishing time: 5 MINUTES | 180°F (82°C)

These spiced plums are an easy way to dress up a store-bought pound cake or sponge cake for a quick and easy dessert. I like spooning the plums over the cake with some of the cooking liquid drizzled on top and topped with some thick Greek-style yogurt or goat's cheese ice cream to balance out the sweetness. The plums are also delicious on their own, chilled. **4 SERVINGS**

6 plums, pitted and quartered
½ cup honey
2 tablespoons water
1 tablespoon freshly squeezed
   lemon juice

1 vanilla bean pod, halved lengthwise,
   seeds scraped out and reserved
1 cinnamon stick

1. Set your sous vide circulator to 180°F (82°C).

2. Combine all the ingredients (including the reserved vanilla bean seeds and pod) in a resealable bag and seal using the water displacement method. Place the sealed bag in the water bath and cook for 45 minutes.

3. Remove the bag from the water bath and remove the vanilla bean pod and cinnamon stick.

4. Serve immediately or store in the cooking liquid in a sealed container in the refrigerator for up to 5 days.

PASSION FRUIT CUSTARD

Prep time: 15 MINUTES | Cook time: 45 MINUTES | Finishing time: 0 MINUTES | 195°F (90.5°C)

My wife is Australian and, growing up, her grandma always had a passion fruit vine in the backyard. Every summer they would cook up a storm with the large haul of passion fruit. This recipe is based on one of the many passion fruit recipes in her family recipe bank. Traditionally they would serve it with a few ladyfingers, using the cookies as a spoon of sorts to dip into the rich custard and provide some crunch. Biscotti or even gingersnap cookies work out well, too. **6 SERVINGS**

12 large egg yolks
1 cup superfine sugar
3 cups heavy (whipping) cream
1 teaspoon vanilla extract

Pulp of 3 fresh passion fruit or 1 (6-ounce) can passion fruit pulp (fresh is always better, if you can find it)

1. Set your sous vide circulator to 195°F (90.5°C).

2. Using an electric mixer on high speed, whisk together the egg yolks and sugar in a bowl until pale and creamy, 1 to 2 minutes.

3. Heat the cream in a saucepan over medium heat to just below boiling point. Be careful not to burn the cream. Add the vanilla and simmer over low heat for 3 to 4 minutes. Remove from the heat and cool for 2 to 3 minutes.

4. When cooled, pour a small amount of the cream mixture into the egg mixture and whisk to combine. (This tempers the egg mixture so it doesn't scramble when you pour in the rest of the hot cream mixture.) Pour the remainder of the cream mixture into the eggs and stir to combine. Add the passion fruit pulp and stir to combine.

5. Pour the custard mixture into six mini mason jars, dividing evenly. Screw on the jar lids to fingertip tightness and place in the water bath, ensuring they are completely submerged. Cook for 45 minutes.

6. Remove the jars from the water bath and allow to cool for 5 minutes. Remove the lids and place the jars in the refrigerator until the custard sets, if serving immediately. Otherwise, store in the refrigerator for up to 3 days and stir before serving.

ORANGE CHOCOLATE POTS DU CRÈME

Prep time: 15 MINUTES, PLUS 5 HOURS COOLING TIME | Cook time: 45 MINUTES
Finishing time: 5 MINUTES | 195°F (90.5°C)

The combination of orange and chocolate is a popular one and can be found in lots of desserts. Whether it is part of a holiday celebration or just a weeknight dessert, this is a winner. I like this dessert dolloped with some lightly sweetened whipped cream. **6 SERVINGS**

6 large egg yolks

1⅓ cups fine white sugar

3 cups heavy (whipping) cream

1 teaspoon vanilla extract

Zest of 1 large orange

⅛ teaspoon orange extract (optional)

⅔ cup chopped semisweet chocolate

2 tablespoons freshly squeezed orange juice

2 tablespoons orange liqueur (optional)

1. Set your sous vide circulator to 195°F (90.5°C).

2. Using an electric mixer on high speed, whisk together the egg yolks and sugar in a bowl until pale and creamy, 1 to 2 minutes.

3. Pour the cream into a saucepan over medium heat and heat to just below boiling point. Be careful not to burn the cream. Add the vanilla, orange zest, and orange extract (if using) to the cream and simmer over low heat for 3 to 4 minutes.

4. Remove the cream mixture from the heat and cool for 2 to 3 minutes. Meanwhile, melt the chocolate using either the double boiler method or in a microwave.

5. Pour a small amount of the cream mixture into the egg mixture and whisk to combine. (This tempers the egg mixture so it doesn't scramble when you pour in the rest of the hot cream mixture.) Pour the remainder of the cream mixture into the eggs and stir to combine. Add the melted chocolate and stir until fully combined. Add the orange juice and orange liqueur (if using), stirring quickly to avoid curdling.

6. Pour the chocolate mixture into six mini mason jars, dividing evenly. Screw on the jar lids to fingertip tightness and place in the water bath, ensuring they are completely submerged. Cook for 45 minutes.

7. Remove the jars from the water bath and allow to cool for 5 minutes. Remove the lids if serving immediately, otherwise store in the refrigerator for up to 3 days.

MAPLE-THYME PEACHES

Prep time: 20 MINUTES | Cook time: 45 MINUTES | Finishing time: 5 MINUTES | 180°F (82°C)

There's nothing better than biting into a ripe juicy peach that you bought from the farmer who picked it a few hours earlier. This recipe has a little Canadian flair with the addition of the maple syrup. Make sure you get real maple syrup; it's a little bit more expensive, but so worth it for the flavor. You can eat these peaches on their own, with ice cream or some whipped mascarpone, or you could make a cobbler topping and sprinkle it over the top of the peaches for some extra crunch. **4 SERVINGS**

6 ripe peaches, pitted and quartered
½ cup good quality Canadian
   maple syrup
2 tablespoons water

1 tablespoon freshly squeezed
   lemon juice
3 fresh thyme sprigs
1 fresh rosemary sprig

1. Set your sous vide circulator to 180°F (82°C).

2. Combine all the ingredients in a food-safe bag and seal using the water displacement method. Place the sealed bag in the water bath and cook for 45 minutes.

3. Remove the bag from the bath and discard the herb sprigs.

4. Serve the peaches immediately or store in the cooking liquid in a sealed container in the refrigerator for up to 5 days.

SOUTHERN DON PEDRO

Prep time: 5 MINUTES | Cook time: 1 HOUR, PLUS COOLING TIME
Finishing time: 5 MINUTES | **165°F (74°C)**

*My parents grew up in South Africa and immigrated to Canada as newlyweds.
As a result I grew up with many classic South African meals served at the dinner
table. One drink in particular caught my attention, mostly because it wasn't one
I was allowed to drink until I was an adult, which just enhanced the fascination.
The drink is called a Don Pedro and is an adult milk shake that is a cross between
dessert and cocktail. Traditionally it is made with a coffee liqueur such as Kahlúa,
however I like it with coffee-infused bourbon.* **LIQUOR SUFFICIENT FOR 17 STANDARD
SHOTS, OTHER INGREDIENTS SUFFICIENT FOR 1 DRINK**

3 cups bourbon
1 cup freshly ground coffee beans
1 cup good quality vanilla ice cream

2 tablespoons heavy (whipping) cream
Chocolate shavings, for garnish

1. Set your sous vide circulator to 165°F (74°C).

2. Divide the bourbon and coffee equally between two large mason jars and
seal firmly.

3. Place the jars in the water bath, ensuring they are completely submerged.
Cook for 1 hour.

4. Remove the jars from the water bath and place in an ice bath to cool. When
cooled, strain the liquid through cheesecloth to remove the coffee grounds.

5. Add 1½ ounces (one standard shot glass) of coffee-infused bourbon, the ice
cream, and cream to a blender. Blend until smooth.

6. Pour into a wineglass and top with chocolate shavings. Serve with a straw.
The remaining bourbon can be stored in a sealed glass container in the refrigerator for up to 4 weeks.

MELANIE'S KISS

Prep time: 5 MINUTES | Cook time: 3 HOURS, PLUS COOLING TIME
Finishing time: 5 MINUTES | 130°F (54°C)

One year my wife spent a lot of time in Frankfurt for work, and the hotel she always stayed at served a cocktail called "Marylin's Kiss," which became her new favorite. She loved it so much that she was determined to try and replicate it at home so she could enjoy it without traveling all the way to Frankfurt! This is the result.

LIQUOR SUFFICIENT FOR 17 STANDARD SHOTS, OTHER INGREDIENTS SUFFICIENT FOR 1 DRINK

3 cups vodka
Zest of 6 large lemons
Dash Sous Vide Simple Syrup (page 190)

6 tablespoons sparkling wine
1 thin twist lemon skin, for garnish

1. Set your sous vide circulator to 130°F (54°C).

2. Divide the vodka and lemon zest equally between two large mason jars and seal firmly.

3. Place the jars in the water bath, ensuring they are fully completely submerged. Cook for 3 hours.

4. Remove the jars from the water bath and place in an ice bath to cool. When cooled, strain the liquid through cheesecloth to remove the lemon zest.

5. Add a dash of Sous Vide Simple Syrup to the base of a champagne glass. Add 1½ ounces (one standard shot glass) of the lemon-infused vodka to the glass and top with the sparkling wine.

6. Place the lemon twist on the rim of the glass and serve. The remaining vodka can be stored in a sealed glass container in the refrigerator for up to 4 weeks.

Tip: Be sure to wash the lemons well to remove all trace chemicals before preparing the zest for this recipe, and also try to avoid including any white pith of the lemon as this will make the end result bitter.

SOMETHING BLUE

Prep time: 5 MINUTES | Cook time: 3 HOURS, PLUS COOLING TIME
Finishing time: 5 MINUTES | 130°F (54°C)

When my wife and I got married we decided we each wanted to select a specialty cocktail for our guests to enjoy. Being a bourbon lover, I went with a classic Old-Fashioned. My wife wanted to create something new for our guests and so she came up with this cocktail. I think the name is spot on. **LIQUOR SUFFICIENT FOR 17 STANDARD SHOTS, OTHER INGREDIENTS SUFFICIENT FOR 1 DRINK**

2 cups fresh blueberries
3 cups vodka
3 ounces old-fashioned lemonade

1 ounce sparkling wine
3 blueberries on a swizzle stick,
  for garnish

1. Set your sous vide circulator to 130°F (54°C).

2. Place 1 cup of blueberries in each of two large mason jars and use the back of a spoon to lightly bruise and break some of the skins on the blueberries. Add the vodka to the mason jars, dividing evenly. Screw on the jar lids to fingertip tightness.

3. Place the jars in the water bath, ensuring they are completely submerged. Cook for 3 hours.

4. Remove the jars from the water bath and place in an ice bath to cool. When cooled, strain the liquid through cheesecloth to remove the blueberries.

5. Place some ice, 1½ ounces (1 standard shot glass) of blueberry-infused vodka, and the lemonade into a cocktail shaker. Shake for 30 seconds.

6. Pour into a martini glass and top with the sparkling wine. Garnish with the blueberry swizzle stick. The remaining vodka can be stored in a sealed glass container in the refrigerator for up to 4 weeks.

VACATION PUNCH

Prep time: 5 MINUTES | Cook time: 2 HOURS, PLUS COOLING TIME
Finishing time: 5 MINUTES | 130°F (54°C)

What flavors could be more representative of a tropical vacation than pineapple, coconut, and lime? No need to pack a suitcase and hop on a flight, just shake up this recipe, sit back, and enjoy—it is a vacation in a glass!

3 cups white rum
1 large pineapple, peeled, cored, and
    chopped into small chunks
1 ounce coconut water

1 ounce pineapple juice
1 tablespoon freshly squeezed lime juice
1 ounce soda water

1. Set your sous vide circulator to 130°F (54°C).

2. Divide the rum and pineapple equally between two large mason jars and screw on the jar lids to fingertip tightness.

3. Place the jars in the water bath, ensuring they are completely submerged. Cook for 2 hours.

4. Remove the jars from the water bath and place in an ice bath to cool. When cooled, use a slotted spoon to remove and reserve the pineapple chunks from the liquid.

5. Place some ice, 1½ ounces (1 standard shot glass) of pineapple-infused rum, coconut water, pineapple juice, and lime juice into a cocktail shaker. Shake for 30 seconds.

6. Add ice to a highball glass and pour the shaken drink over. Garnish with a few chunks of reserved pineapple and top with the soda water. The remaining rum can be stored in a sealed glass container in the refrigerator for up to 4 weeks.

ORANGE OLD-FASHIONED

Prep time: 5 MINUTES | Cook time: 2 HOURS, PLUS COOLING TIME
Finishing time: 5 MINUTES | **165°F (74°C)**

I love a good old-fashioned and wherever I go, if there is one on the menu I feel compelled to order it. As a result, I have tasted all sorts of variations of this drink and to be honest, the original is still the best. This recipe is a close second. I think the enhanced orange flavor of the bourbon really complements the spicy sweetness of bourbon. **LIQUOR SUFFICIENT FOR 17 STANDARD SHOTS, OTHER INGREDIENTS SUFFICIENT FOR 1 DRINK**

3 cups bourbon
Zest of 6 large oranges
2 teaspoons Sous Vide Simple Syrup
(page 190)

3 dashes Angostura bitters
1 orange, for garnish

1.   Set your sous vide circulator to 165°F (74°C).

2.   Divide the bourbon and orange zest equally between two large mason jars and screw on the jar lids to fingertip tightness.

3.   Place the jars in the water bath, ensuring they are completely submerged. Cook for 2 hours.

4.   Remove the jars from the water bath and place in an ice bath to cool. When cooled, strain the liquid through cheesecloth to remove the orange zest.

5.   Add the Sous Vide Simple Syrup, bitters, 1½ ounces (1 standard shot glass) of bourbon, and a few ice cubes to a rocks glass. Stir for 30 seconds.

6.   Peel a wide but thin strip of orange peel over the glass so that the essential oils spray over the drink. Add the peel to the glass as a garnish and serve. The remaining bourbon can be stored in a sealed glass container in the refrigerator for up to 4 weeks.

WATERMELON MOJITO

Prep time: 5 MINUTES | Cook time: 2 HOURS, PLUS COOLING TIME
Finishing time: 5 MINUTES | 130°F (54°C)

This is a really great cocktail to drink on the dock on a summer afternoon or to serve as part of a summer barbecue. The watermelon and mint combination is refreshing, and serving this in a tall glass with ice ensures the drink is not too alcohol forward for a summer afternoon and keeps everyone cool. **LIQUOR SUFFICIENT FOR 17 STANDARD SHOTS, OTHER INGREDIENTS SUFFICIENT FOR 1 DRINK**

2 cups cubed fresh watermelon

3 cups white rum

5 to 8 fresh mint leaves, torn roughly

1 tablespoon Sous Vide Simple Syrup (page 190)

1 tablespoon freshly squeezed lime juice

2 ounces soda water

1. Set your sous vide circulator to 130°F (54°C).

2. Place 1 cup of watermelon in each of two large mason jars. Divide the rum evenly between the mason jars. Screw on the jar lids to fingertip tightness.

3. Place the jars in the water bath, ensuring they are completely submerged. Cook for 2 hours.

4. Remove the jars from the water bath and place in an ice bath to cool. When cooled, strain the liquid through cheesecloth to remove the watermelon. Reserve the watermelon for garnish.

5. Place the mint leaves, Sous Vide Simple Syrup, and lime juice in a highball glass and use either a muddler or the back of a spoon to crush the mint leaves.

6. Add some ice cubes and 1½ ounces (1 standard shot glass) of watermelon-infused rum, then top with the soda water.

7. Thread a few cubes of reserved watermelon on a swizzle stick and use to garnish the drink. Serve immediately. The remaining rum can be stored in a sealed glass container in the refrigerator for up to 4 weeks.

MULLED WINE

Prep time: 15 MINUTES | Cook time: 1 HOUR, 30 MINUTES | **165°F (74°C)**

With the temperature dipping below zero during the holiday season in Canada, mulled wine is a staple at holiday parties. There is nothing more enticing than stepping out of the snow into a friend's home and being handed a steaming hot mug of mulled wine. It certainly helps the festive cheer blossom! **4 TO 5 SERVINGS**

3 cups red wine

1 large orange, sliced

2 cinnamon sticks

1 cup plus 2 tablespoons brandy

¼ cup Sous Vide Simple Syrup (page 190)

10 whole cloves

4 star anise

2 vanilla pods, halved lengthwise

1. Set your sous vide circulator to 165°F (74°C).

2. Divide all ingredients evenly between two large mason jars and screw on the jar lids to fingertip tightness.

3. Place the jars in the water bath, ensuring they are completely submerged. Cook for 1 hour, 30 minutes.

4. Remove the jars from the water bath and allow to cool for 3 minutes. Pour into ceramic mugs and serve immediately.

STRAWBERRY MARGARITA

Prep time: 5 MINUTES | Cook time: 2 HOURS, PLUS COOLING TIME
Finishing time: 5 MINUTES | 130°F (54°C)

This is a sweet twist on a classic margarita. I find that the sweetness of the straw-berry works really well with the fiery herbaceous flavor of tequila. The recipe below is for a margarita on the rocks, but if you prefer a slushy-style margarita, the ingre-dients can easily be combined in a blender for an icy treat. **LIQUOR SUFFICIENT FOR 17 STANDARD SHOTS, OTHER INGREDIENTS SUFFICIENT FOR 1 DRINK**

2 cups fresh strawberries, stemmed
 and halved
3 cups tequila
1 tablespoon Sous Vide Simple
 Syrup (page 190)

2 tablespoons freshly squeezed lime juice
2 tablespoons orange liqueur like Triple
 Sec or Cointreau
Lime wedge, for garnish
Rock salt, for garnish

1. Set your sous vide circulator to 130°F (54°C).

2. Place 1 cup of strawberries in each of two large mason jars. Divide the tequila evenly between the mason jars. Screw on the lids to fingertip tightness.

3. Place the jars in the water bath, ensuring they are completely submerged. Cook for 2 hours.

4. Remove the jars from the water bath and place in an ice bath to cool. When cooled, strain the liquid through cheesecloth to remove the strawberries. Reserve the strawberries and blend them in a blender until smooth to make the straw-berry purée.

5. Place some ice, 2 ounces of strawberry-infused tequila, the Sous Vide Simple Syrup, lime juice, orange liqueur, and 1 teaspoon of strawberry purée into a cock-tail shaker. Shake for 30 seconds.

6. Rub the lime wedge around the rim of a rock glass and dip the glass rim into a saucer spread with rock salt.

7. Pour the shaker contents into the prepared glass and serve immediately. The remaining rum can be stored in a sealed glass container in the refrigerator for up to 4 weeks.

ESPRESSO MARTINI

Prep time: 5 MINUTES | Cook time: 1 HOUR, PLUS COOLING TIME
Finishing time: 5 MINUTES | 130°F (54°C)

An espresso martini is a great option to serve instead of dessert at a dinner party. The coffee bean infusion is like a wake-up call, the Sous Vide Simple Syrup (page 190) and coffee give you some sweetness, and the vodka provides a little kick. I once had one sprinkled with feuilletine flakes and it was one of the most delicious things I have ever tasted! For those of us who don't have a bag of feuilletine flakes handy, a few espresso beans do the trick. **LIQUOR SUFFICIENT FOR 17 STANDARD SHOTS, OTHER INGREDIENTS SUFFICIENT FOR 1 DRINK**

3 cups vodka
1 cup freshly ground espresso beans
1 teaspoon Sous Vide Simple Syrup
  (page 190)

2 tablespoons coffee liqueur
2 tablespoons fresh brewed espresso
3 espresso beans, for garnish

1. Set your sous vide circulator to 130°F (54°C).

2. Divide the vodka and ground espresso beans equally between two large mason jars. Screw on the jar lids to fingertip tightness.

3. Place the jars in the water bath, ensuring they are completely submerged. Cook for 1 hour.

4. Remove the jars from the water bath and place in an ice bath to cool. When cooled, strain the liquid through cheesecloth to remove the coffee grounds.

5. Place some ice, 1½ ounces (1 standard shot glass) of coffee-infused vodka, the Sous Vide Simple Syrup, coffee liqueur, and espresso in a cocktail shaker. Shake for 30 seconds.

6. Pour into a martini glass and garnish with the espresso beans. Serve immediately. The remaining vodka can be stored in a sealed glass container in the refrigerator for up to 4 weeks.

BERRY GIN *and* TONIC

Prep time: 5 MINUTES | Cook time: 2 HOURS, PLUS COOLING TIME
Finishing time: 5 MINUTES | 130°F (54°C)

This recipe is a little bit of a change from the run-of-the-mill gin and tonic. It is very versatile—if you aren't a fan of blackberries you could substitute any other type of berry and get an equally delicious and refreshing drink. **LIQUOR SUFFICIENT FOR 17 STANDARD SHOTS, OTHER INGREDIENTS SUFFICIENT FOR 1 DRINK**

3 cups gin
2 cups fresh blackberries, plus more
   for garnish

1 teaspoon Sous Vide Simple
   Syrup (page 190)
1 teaspoon freshly squeezed lemon juice
⅓ cup tonic water

1.   Set your sous vide circulator to 130°F (54°C).

2.   Divide the gin and blackberries equally between two large mason jars. Screw on the jar lids to fingertip tightness.

3.   Place the jars in the water bath, ensuring they are completely submerged. Cook for 2 hours.

4.   Remove the jars from the water bath and place in an ice bath to cool. When cooled, use a slotted spoon to remove and discard the blackberries from the gin.

5.   Place some ice, 1½ ounces (1 standard shot glass) of blackberry-infused gin, the Sous Vide Simple Syrup, and lemon juice in a highball glass and stir for 30 seconds.

6.   Top with the tonic water and garnish with a few blackberries. Serve immediately. The remaining gin can be stored in a sealed glass container in the refrigerator for up to 4 weeks.

chapter eight

The Kitchen Sink

There are other ways to use sous vide apart from cooking proteins and vegetables. You can use your sous vide circulator to create other components that amplify your meals, add depth to your oils, and even pickle vegetables. You can get creative and also create things that you use often. I'd love to hear from readers on the ways you use your sous vide outside of the box.

BREAD AND BUTTER PICKLES (page 193)

EASY DULCE DE LECHE

Prep time: 1 MINUTE | Cook time: 14 HOURS | Finishing time: 0 MINUTES | **185°F (85°C)**

Dulce de leche is a fancy way of saying caramel, and it is very popular in Latin American countries. It's created by cooking condensed milk for a long period of time, which results in the thickening and caramelizing of the milk. This recipe can easily be scaled up or down depending on what you intend to use it for. It will keep in the refrigerator for up to 3 weeks, although it always seems to disappear much faster in my house! It is very versatile and can be used as a simple topping for ice cream, a flavor layer in a cake, or in other more complex recipes. **2 CUPS**

2 cups sweetened condensed milk

1. Set your sous vide circulator to 185°F (85°C).

2. Pour 1 can of condensed milk into each of two mason jars.

3. Screw on the jar lids to fingertip tightness, and place in the water bath, ensuring they are completely submerged. Cover the container to minimize water evaporation.

4. Cook for 14 hours, checking periodically to ensure the water level does not drop.

5. Remove the jars from the water bath and store in the refrigerator for up to 3 weeks.

SOUS VIDE SIMPLE SYRUP

Prep time: 2 MINUTES | Cook time: 2 HOURS | Finishing time: 0 MINUTES | **140°F (60°C)**

I always like to have a good stock of simple syrup on hand as it is a great starting point for sweet sauces and flavor infusions for cocktails. I usually cook up a large batch to have it at the ready in the refrigerator, where it will store for up to 5 weeks. This recipe can be scaled up or down as needed, and as the name suggests, it doesn't get much simpler than this. **3 CUPS**

3 cups water
3 cups superfine sugar

1. Set your sous vide circulator to 140°F (60°C).

2. Combine the water and sugar in a large food-safe bag (or a few smaller bags) and seal using the water displacement method. Place the sealed bag in the water bath and cook for 2 hours.

3. Remove the bag from the water bath and pour the syrup into a sealable jar or bottle. Store in the refrigerator for up to 5 weeks.

BACON MAPLE THYME BUTTER

Prep time: 5 MINUTES | Cook time: 2 HOURS, PLUS COOLING TIME
Finishing time: 15 MINUTES | 195°F (90.5°C)

This butter stores well for a few weeks in the refrigerator and is great served over a delicious sous vide steak! **1 STICK OF BUTTER**

8 tablespoons (1 stick) butter, cut into
   small chunks
4 tablespoons maple syrup

3 fresh thyme sprigs
5 bacon slices

1.   Set your sous vide circulator to 195°F (90.5°C) and heat the oven to 400°F (205°C).

2.   Place the butter, maple syrup, and thyme in a large mason jar and screw on the jar lid to fingertip tightness.

3.   Place the jar in the water bath, ensuring it is completely submerged. Cook for 2 hours.

4.   Meanwhile, lay the bacon on an aluminum foil–lined baking sheet. Bake in the oven for 20 minutes. Remove from the oven and cool. When cool, cut the bacon into small pieces.

5.   Remove the jar from the water bath and cool for 2 minutes. Strain the liquid through cheesecloth to remove the thyme.

6.   Mix the bacon pieces into the liquid butter and stir to combine, then pour the liquid butter into a sealable glass container and refrigerate for at least 1 hour to solidify. Use as needed. The butter will store in the refrigerator for up to 4 weeks.

MIXED BERRY COULIS

Prep time: 5 MINUTES | Cook time: 45 MINUTES | Finishing time: 0 MINUTES | 180°F (82°C)

This mixed berry coulis is very versatile. It can be served in a smooth and refined form or a more rustic homely form. You can also change up the berry composition to suit your personal taste; just be sure to include 2 cups of fruit in total. **1 TO 2 CUPS**

½ cup fresh strawberries, chopped
½ cup fresh raspberries
½ cup fresh blackberries
½ cup fresh blueberries

⅓ cup Sous Vide Simple Syrup (page 190) or ½ cup superfine white sugar
3 tablespoons freshly squeezed lemon juice

1.   Set your sous vide circulator to 180°F (82°C).

2.   Combine the berries, Sous Vide Simple Syrup, and lemon juice in a large food-safe bag and seal using the water displacement method. Place the sealed bag in the water bath and cook for 45 minutes.

3.   Remove the bag from the water bath. If you would like your coulis smooth, place the contents into a blender and blend until smooth. Strain the mixture through a fine mesh strainer to remove seeds and then serve. If you prefer a more rustic texture, remove the contents from the bag and serve as is. The coulis can be stored in a sealed container in the refrigerator for up to 1 week.

BREAD *and* BUTTER PICKLES

Prep time: 5 MINUTES | Cook time: 30 MINUTES | Finishing time: 1 MINUTE | 180°F (82.2°C)

I've used sous vide techniques to pickle radishes as part of a Thomas Keller cooking endeavor. His book Under Pressure is one of the most beautifully photographed cookbooks, but it is also well beyond the reach of most home cooks. Some of the components of his extravagant dishes can be made at home, but for the most part, the book punches above the average weight class. The pickling methods—applying heat and vinegar to vegetables—however, do apply and this 30-minute recipe is an example of that. Be forewarned: These are addictive. **24 PICKLES**

2 English cucumbers
1 cup champagne vinegar
½ cup powdered sugar

2 teaspoons kosher salt
1 tablespoon Bahārāt spice
½ white onion, chopped

1. Set your sous vide circulator to 180°F (82.2°C).

2. Use a sharp knife to cut the cucumber into ½-inch slices, 3 inches long.

3. Combine the vinegar, powdered sugar, salt, and Bahārāt spice in a small bowl and whisk to mix.

4. Add the cucumbers to a medium mason jar, followed by the onion and spiced vinegar mix. Screw on the jar lid to fingertip tightness.

5. Place the jar in the water bath, ensuring it is fully submerged, and cook for 30 minutes.

6. When the cucumbers are cooked, remove the lid and let it sit for 5 minutes. Remove the pickles and place on a paper towel–lined plate to cool.

7. Rinse the jar and place the pickles and cooking liquid inside. The pickles will keep in a lidded jar for up to 30 days.

LEMON-INFUSED WHITE BALSAMIC VINEGAR

Prep time: 5 MINUTES | Cook time: 3 HOURS | Finishing time: 1 MINUTE | 153°F (67.2°C)

Mel and I purchased lemon white balsamic vinegar at the Toronto Food Fest in 2016. We added it to everything—salads, dipping for breads, marinades for fish, you name it. Knowing that sous vide is a fantastic method for infusions, this recipe was a no-brainer. White balsamic vinegar can be tough to find, but if you have a local food market (or a food festival, in our case), you can find it with a little bit of sleuthing. **2 CUPS**

2 fresh lemons, peeled
4 cups white balsamic vinegar

1. Set your sous vide circulator to 153°F (67.2°C).

2. Use a paring knife to peel the lemons, making sure to limit the amount of rind, and set aside.

3. Twist the lemons to produce oils, and place in a mason jar. Pour the vinegar into the jar and screw on the jar lid to fingertip tightness. Place the jar in the water bath, ensuring it is completely submerged. Cook for 3 hours.

4. Remove the lid and pour the vinegar solution through a mesh strainer, transfer to a bottle, and store for up to 60 days.

CHILI OIL

Prep time: 5 MINUTES | Cook time: 3 HOURS | Finishing time: 1 MINUTE | **135°F (57.2°C)**

Most pizza and pasta joints worth a shake will offer you chili oil to hit your dish with some extra heat. They vary in terms of heat, and so does this recipe. As you might expect, the longer the chiles remain in contact with the oil, the more heat they will impart. This base recipe calls for you to remove the chiles after cooking, but if you want things hotter, do one of two things: Use a hotter pepper (Scotch bonnet for heroes), or stick with the recipe and remove the peppers from the bottle once you've hit your maximum tolerance. The heat of the sous vide will release the spice from the peppers so be aware that jalapeño peppers, under heat, add some decent fire to the oil. **2 CUPS**

2 fresh habanero peppers

2 fresh jalapeño peppers

4 cups light olive oil

2 teaspoon dried red pepper flakes

1. Set your sous vide circulator to 135°F (57.2°C).

2. Use a paring knife to quarter the habanero and jalapeño peppers.

3. Combine the peppers, olive oil, and red pepper flakes in a medium mason jar and screw on the jar lid to fingertip tightness. Place the jar in the water bath, ensuring it is completely submerged, and cook for 3 hours.

4. Transfer the mason jar to ice bath and let cool.

5. Pour the oil through a mesh strainer, transfer to a bottle, and refrigerate for up to 60 days.

PICKLED CAESAR GREEN BEANS

Prep time: 5 MINUTES | Cook time: 15 MINUTES | Finishing time: 1 MINUTE | **190°F (87.8°C)**

If you're a fan of Bloody Marys or Caesars, you'll know that a key ingredient that makes the drink is the spicy pickled green bean. Now you can quickly make your own for a weekend brunch or a mid-afternoon cocktail. **24 BEANS**

¾ **pound green beans, trimmed**
2 **fresh jalapeño peppers**
⅔ **cup champagne vinegar**
⅔ **cup water**

3 **tablespoons sugar**
1 **tablespoon sea salt**
2 **teaspoons red pepper flakes**

1. Set your sous vide circulator to 190°F (87.8°C).

2. Use a sharp knife to cut the beans to a length that will fit into a mason jar.

3. Cut the jalapeños widthwise into 1-inch strips.

4. Combine the vinegar, water, sugar, salt, red pepper flakes, and jalapeños in a small saucepan over medium-high heat and bring to a boil.

5. Transfer the beans and vinegar solution to the mason jar and screw on the jar lid to fingertip tightness. Place the jar in the water bath, ensuring it is completely submerged, and cook for 15 minutes.

6. When the beans are cooked, transfer the mason jar to an ice bath for 15 minutes.

7. Serve immediately or store in the jar for up to 7 days.

FOUR-HERB BUTTER

Prep time: 10 MINUTES | Cook time: 2 HOURS | Finishing time: 5 MINUTES | 195°F (90.5°C)

Herbed butter is an amazing way to add flavor to breads, serve with fresh fish, use for frying, and finishing poultry and beef, just to name a few. This is a really simple recipe and will stay preserved in the refrigerator for weeks. ½ **CUP**

8 tablespoons (1 stick) unsalted butter
1 tablespoon parsley, chopped
   finely, divided
½ tablespoon fresh chives, chopped finely

½ tablespoon fresh basil leaves,
   chopped finely
½ tablespoon fresh thyme leaves,
   chopped finely

1. Set your sous vide circulator to 195°F (90.5°C).

2. Cut the butter into small chunks and put in a mason jar, followed by the parsley, chives, basil, and thyme. Screw on the jar lid to fingertip tightness. Place the jar in the water bath, ensuring it is completely submerged, and cook for 2 hours.

3. Once cooking is complete, remove the jar from the water bath and set aside to cool for 10 minutes.

4. Remove the lid and strain the butter through cheesecloth into a clean mason jar. Discard the herbs.

5. Cover the jar with a lid and store in the refrigerator for up to 30 days.

cooking times and temperatures

This book outlines cooking temperatures and times based on personal preference. The information here will provide more detail on the ranges of doneness and cooking times (both minimum and maximum times) to help you build your own recipes. These ranges show the minimum times that certain foods need to be cooked in order to be safe to eat. You will also see that some foods can cook for longer than the recipes in this book state, giving you some flexibility in timing.

Beef

Steaks (tender cuts) should be cooked for at least 1 hour and typically no more than 3 hours.

> **RARE:** 129°F/54°C—1 to 2 hours
> **MEDIUM:** 136°F/58°C—1 to 2 hours
> **WELL DONE:** 154°F/68°C—1 to 2 hours

Roasts (prime rib, rib roast) should be cooked for longer to help break down connective tissue, between 5 and 16 hours.

> **RARE:** 133°F/56°C—5 to 16 hours
> **MEDIUM:** 140°F/60°C—6 to 12 hours
> **WELL DONE:** 158°F/70°C—5 to 10 hours

Tough Cuts (brisket, chuck roasts, eye of round) need to be cooked for at least 8 hours and up to 48 hours.

> **RARE:** 131°F/55°C—24 to 48 hours
> **MEDIUM:** 149°F/65°C—20 to 24 hours
> **WELL DONE:** 185°F/85°C—8 to 16 hours

Pork

Chops (loins) should be cooked for at least 1 hour and typically no more than 3 hours.

    RARE: 136°F/58°C—1 to 3 hours
    MEDIUM: 144°F/62°C—1 to 3 hours
    WELL DONE: 158°F/70°C—1 to 3 hours

Roasts (prime roasts) should be cooked for longer to help break down connective tissue, between 3 and 6 hours.

    RARE: 136°F/58°C—3 to 6 hours
    MEDIUM: 144°F/62°C—3 to 4 hours
    WELL DONE: 158°F/70°C—3 to 3 hours, 30 minutes

Tough Cuts (belly, shoulder) need to be cooked for at least 8 hours and up to 24 hours.

    RARE: 140°F/60°C—8 to 24 hours
    MEDIUM: 154°F/68°C—8 to 24 hours
    WELL DONE: 185°F/85°C—8 to 16 hours

Chicken

Light Meat (breasts) should be cooked for at least 1 hour and no more than 3 hours.

    SOFT AND JUICY: 140°F/60°C—1 to 3 hours
    TENDER AND JUICY: 149°F/65°C—1 to 3 hours
    WELL DONE: 167°F/75°C—1 to 3 hours

Dark Meat (legs, thighs) should be cooked for at least 1 hour and no more than 5 hours.

    TENDER AND JUICY: 149°F/65°C—1 to 5 hours
    FALL OFF BONE: 167°F/75°C—1 to 5 hours

Fish

All **fillets** should be cooked for at least 45 minutes and no more than 1 hour, 15 minutes.

TENDER AND SOFT: 104°F/40°C—45 minutes to 1 hour, 15 minutes
TENDER AND FLAKY: 122°F/50°C—45 minutes to 1 hour, 15 minutes
WELL DONE: 131°F/55°C—45 minutes to 1 hour, 15 minutes

Vegetables

Green vegetables differ from root vegetables and cooking times can range from 10 minutes to 3 hours, but all temperatures remain the same.

GREEN VEGETABLES: 185°F/85°C—10 to 20 minutes
GOURDS: 185°F/85°C—1 to 3 hours
ROOT VEGETABLES AND POTATOES: 185°F/85°C—2 to 3 hours

Fruits

Fruit cooking time and temperatures vary based on what you want to make—poached or purée.

WARM AND RIPE: 154°F/68°C—1 hour, 45 minutes to 2 hours, 30 minutes
SOFT: 185°F/85°C—30 minutes to 1 hour, 30 minutes

resources

There are several great resources available to help you develop your sous vide cooking techniques and style. I looked to many of these when I first started, and I still gather inspiration and ideas from them today.

ChefSteps—ChefSteps.com
This is a beautiful site for ideas, recipes, and techniques. Their videos are simple, engaging, and often funny. They are the maker of the Joule sous vide circulator and bring a combination of over 50 chefs, scientists, photographers, writers, and engineers to focus on "cooking smarter" by using technology and techniques to help people get more out of cooking.

Serious Eats—SeriousEats.com
Kenji López-Alt has been at the forefront of sous vide cooking for many years. He creates easy-to-follow and in-depth guides on how to tackle some of the most common and most challenging foods. Serious Eats creates content very regularly and is a great place to get some new ideas and perfect the details of award-winning dishes. Here is the link to seasoning a cast iron pan: www.seriouseats.com/2010/06/how-to-buy-season-clean-maintain-cast-iron-pans.html.

Sous Vide Subreddit—reddit.com/r/SousVide
Some of you are certainly Redditors, I know that for sure. For those of you who don't know what Reddit is, or haven't used it before, it is a fantastic source of information on any and every topic including sous vide cooking. At the time of writing, this subreddit (topic) had over 34,000 subscribers. You'll find posts every hour or so of Redditors posting pictures, questions, stories, failures, and successes, and the community will comment and discuss to provide thanks, guidance, feedback, recipes, and other interactions. It's very informal compared to the structure of Serious Eats and ChefSteps, but the content is updated all day and night so there is lots to read at any given time.

recipe index

index

acknowledgments

Special thanks to my wife, Melanie, who helped open my eyes to a whole new world of cooking, tastes everything I make even if she's not hungry, and has supported me since day one, especially while writing this book during a very busy time in our lives.

I'd also like to thank my mother, Rose, and my two sisters, Tarryn and Nicki, for jumping on the sous vide bandwagon with me and for sharing their stories, tips, and ideas and letting me cook a sous vide Christmas dinner. Thanks to my dad, Chris, and my two brothers-in-law, Ross and Jason, for providing me with real-time feedback as they eat my sous vide creations.

To my friends Martin and Shannon and Karl, Dona, and Tyler, who have taken care of my beloved Canadian sous vide collection since I left for London: Thanks for putting them to good use and sharing all of your cooks with me. I'm glad you caught the bug and enjoy this as much as I do.

To my fellow Sous Vide Redditors: Keep on cooking and sharing those ideas. It's a great community and grows every day as sous vide gains popularity. Upvotes to all of you.

about the author

Carey Copeling is a self-taught Tech Chef based out of London, England. Originally from Oakville, Ontario, Canada, he began cooking with sous vide methods in 2015 and since then has run several private in-home cooking demonstrations within the Toronto area to demonstrate to guests the secrets and tricks to mastering sous vide cooking. He is a member of the Anova Culinary Hall of Fame, inducted in the initial class in March 2017. Follow him on Instagram @sous.veats.

CPSIA information can be obtained
at www.ICGtesting.com
Printed in the USA
LVHW02s1132240918
590936LV00003BA/3/P

9 781623 159818